S0-AQI-723

STEALING FIRE

STEALING FIRE:

The Atomic Bomb As Symbolic Body

by Peter C. Reynolds

ISBN 0-9629261-0-8
Printed in the United States of America
First Edition

Published by Iconic Anthropology Press,
imprint of Iconic Anthropology Inc.
This book can be purchased directly from the publisher.
See the last page for ordering information.

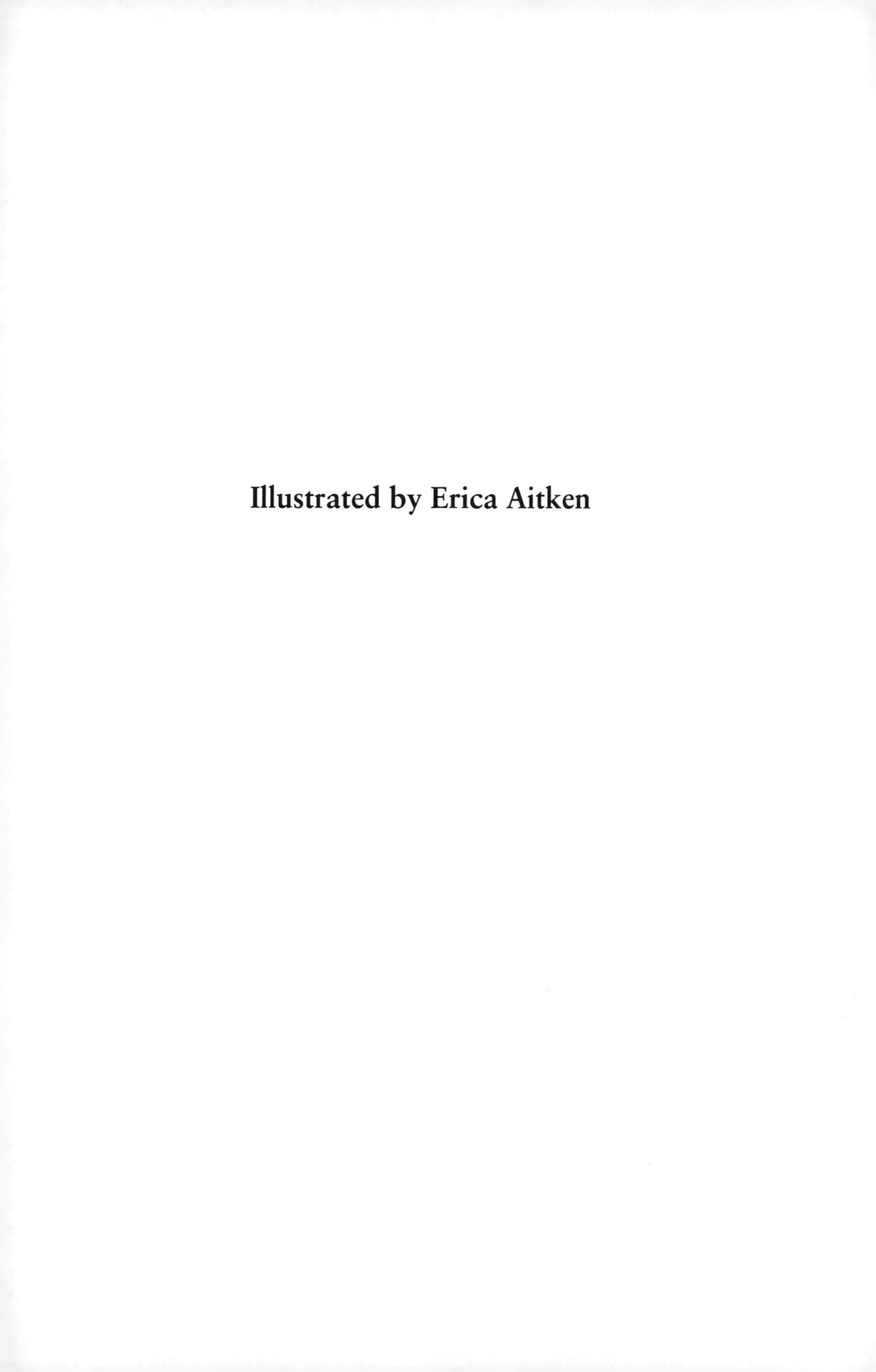

Illustrated by Erica Aitken

Dedication

This book is dedicated to my parents,
Walter and Bernadette,
for their common sense and love of good books.

TABLE OF CONTENTS

INTRODUCTION

In a dozen epic television shows, the sunset shot of the stalking hunter dissolves to the dawn of civilization, but this picture of the past no longer has the power to inspire or inform. To the technocrat and the industrialist, technology is a collection of tools and techniques that defines a society's rank on a uniform scale of technical and economic development, but to the anthropologist *technology* reflects a shared image of the human species and its relationship to nature, what I call the *symbolic body*. In the new approach to human history presented in this book, the differences between "primitive" and "advanced" technologies, far from being facts of history demonstrated by science, are *projections* of the industrial world view onto the human past, and even such commonplace examples of progressive technological development as the steam engine can be shown, on closer analysis, to be highly selective readings of historical events.

In Chapter 1, Pumping Salmon, I show that the industrial symbolic body underlies our present ecological difficulties, for it assumes that "nature" is part of the primitive past, "science" part of the future—and "progress" the transformation of nature and its replacement by machines. But the output of the contemporary industrial system is not progress in the original, humanitarian sense of the term, namely an improvement in the quality of life, but what I call the *one-two punch:* the destruction of natural processes on the one hand and their replacement by technocratic constructions on the other.

This technocratic process of transformation, so often portrayed as the rational application of scientific principles, is far more easily explained by what anthropologists call *myths:* stories of past events, assumed to be true by members of a community, which justify the structure of human institutions and define appropriate action. In industrial society, however, myth is regarded as part of the primitive past that is superseded by science, so the technocracy must disguise its own mythology as seemingly objective descriptions of historical processes and natural events. But if one examines the history of technical innovation, the social organization of the modern technocracy, and the historical context of scientific advances, then it is easy to show that industrial technology is organized by a constellation of myths not very different from those recorded by field anthropologists in so-called primitive societies. The core myths of industrialism—Nebuchadnezzar's Dream, Bloody Stumps/Exploding Stars, The Lone Galileo, The Fire Twins, and The Last Days of Marcus Karenin—define a collective image of the human species and its relationship to nature; and together they comprise an integrated vision of the cosmos, which is mapped to stellar evolution on one end of the time scale and to individual lives on the other. This culturally constructed image of the human species defines the industrial symbolic body, and it organizes cooperative construction and economic activity, reshaping nature into images of itself.

In the myth of Nebuchadnezzar's Dream, for example, primitive technologies are equated with the lower parts of the human body and advanced technologies with the head. In Chapter 2, I review the history of technology from the viewpoint of historians and archaeologists showing that technical progress is more easily explained as a projection of this imagery of head and tail than as a product of rationalist, utilitarian, and materialist processes. Moreover, when these mythic images are intellectualized and projected back to us through museum displays and educational television, we end up with the familiar arrow of human evolution in which small-brained bipeds with crude stone tools evolve over time into erect, intellectual giants, floating free of the earth in spaceships.

In Chapter 3, through analysis of the movie *Star Wars,* I examine the industrial body image, showing that it is based on binary oppositions between nature/culture and male/female. The industrial symbolic body is created and displayed through ritual transformations of the biological body and through the construction of artifacts that incorporate somatic imagery, in which the head, equated with males, is transformed into exploding stars, leaving behind a bloody stump equated with the female. However, as this symbolic system contradicts the technocracy's official ideology of science, materialism, and social equality, it is never acknowledged verbally but has to be encoded in the form of science fiction or unconsciously incorporated into industrial artifacts and techniques.

Chapter 4 demonstrates that science itself is a mythological enterprise that builds on the technocratic body image by replicating a charter myth called the tale of The Lone Galileo. Its primary function is to transform nature into images of exploding stars, and far from providing an objective picture of the natural world, its theories of nature are a surface structure phenomenon which demonstrate the power of the underlying myths. Moreover, the myth of The Lone Galileo encodes a structural opposition between science and organized religion, which from an anthropological perspective is a mythical inversion of Judeo-Christian theology. This indicates that science, far from being the secular institution that it claims to be, is a religion in a strict sense of the term, for it references, albeit covertly, theological propositions; and it offers to its practitioners a personal and collective immortality. By examining key events in the history of science, particularly Darwin's theory of evolution, I show that the establishment of scientific theories is more satisfactorily explained by its covert atheological agenda than by empirical discoveries or theoretical advances.

In Chapter 5, The Twins Who Bring the Fire, I show that high-tech innovations, far from being logical consequences of scientific theory, are better explained as projections of the symbolic body and its associated imagery. Drawing on official histories and technical descriptions of Anglo-American technology during World War II, I question the concept of a scientific discovery, redefining it in purely anthropological terms as *an image of nature that validates the charter mythology of science itself.* Moreover, two of the most famous and characteristic "discoveries" of this period, namely, penicillin and the nuclear chain reaction, are not primarily derived from scientific theory but from technocratic mythology. This interpretation is confirmed by the architecture of the atomic bombs themselves and by the names, events, and settings of the Manhattan Project. Moreover, penicillin and atomic bombs share the same symbolic imagery, which indicates that biology and physics are complementary parts of the symbolic body, joined in reciprocal acts of mutilation and prosthesis—yet another one-two punch.

This one-two punch of physics and biology became institutionally embodied during the Second World War, and it is itself chartered by a myth that I call The Last Days of Marcus Karenin. This constellation of images, explicated in Chapter 6, is the most subtle of the technocratic myths—and by far the most dangerous, for it charters the transformation of the 20th-century technocracy from a covert to an overt religion with apocalyptic tendencies. Symbolic analysis of contemporary science journalism and science fiction indicates that the technocracy, long an ally of religious skepticism and political liberalism, is altering its own structural relationship to the larger society, replacing the vague vision of a future technological paradise with a promise of literal immortality through bio-engineering and a global economy of synthesized body parts.

The epilog, The Color Green, points out the inability of contemporary reform movements to deal with these developments, for the former typically are based on exactly the sorts of liberal and progressivist assumptions that are being made irrelevant by the dominance of technocratic institutions. Moreover, they attack the symptoms of the modern symbolic body, such as nuclear weapons, habitat destruction, and social disintegration, without coming to terms with the deeper mythological premises that make such actions seem rational and progressive. To deal effectively with contemporary social and political problems, we need to reestablish a more humane concept of the symbolic body itself by discrediting the intellectual premises on which industrial mythology is based. This critical stance begins within anthropology itself by questioning the concept of a single evolutionary time line, validated by physics—the absolute chronology—which ranks every event in the history of the planet on a single scale of before and after. Symbolic analysis indicates that this picture of the past is based on the systematic exclusion of physical processes from scientific theory: either those that do not fit the androcentric imagery of the technocratic mythology or those that fail to reinforce the covert atheological agenda of science itself. That is to say, physics' own concept of physical reality is flawed because it does not take account of the mythical biases that affect its own observations.

Therefore, I suggest that the geocentric system of reference rejected by Galileo is in fact the intellectual perspective most relevant to an ecologically balanced society.

This is the Second Coming of Prometheus,
unbound at last after some half a million years,
bringing down a fire from the original flame
that had lighted the stars from the beginning.

William L. Laurence, the only journalist
allowed to witness the first atomic bomb test.

From Laurence, William L. 1959. *Men and Atoms: The Discovery, the Uses, and the Future of Atomic Energy*. New York: Simon and Schuster; p. 7.

PUMPING SALMON

The Reciprocal Relation Between Technology and Myth

A few years ago, my wife and I stopped to look at a large and imposing dam built on a small stream in northern California, at Warm Springs in rural Sonoma County. At the dam site there is a fish hatchery, run by the California Department of Fish and Game, to help replenish the populations of sea-run salmon and steelhead trout that used the stream as a spawning ground before the dam was built. A video at the visitor center shows how gravid trout and salmon, swimming upstream to lay their eggs, are captured at the spillway and put into special holding tanks. They are then anesthetized and lifted out of the water, where teams of biological technicians remove the eggs from the females and fertilize them with sperm from the males. Next, the fertilized eggs are deposited in special incubation trays, each carefully marked with time, date, and batch number, through which water is continually pumped. Banks of pumps, aerators, purifiers, and automated monitoring equipment exactly reproduce the salinity, acidity, temperature, and oxygen content of fresh stream water flowing over a gravel bar. When the eggs hatch, the fish begin a carefully orchestrated migration through a succession of nursery tanks until they are large enough to journey to the sea. At that time, they are sucked into refrigerated tank trucks and driven downstream, past the dam and the biological hazards, to be released nearer the ocean.

Significantly, most of the hazards faced by migrating salmon are themselves products of industrial society. The Warm Springs Dam visitor center contains a large roulette wheel that illustrates the probabilities of a salmon reaching maturity and returning home to spawn. Each spoke of the wheel is labeled with a risk factor: logging activity upstream washes mud over the gravel nesting grounds; mine tailings pollute the stream; agricultural irrigation lowers the water level; and so on. Nearly all the risk factors on this roulette wheel are due to human activities and not to natural catastrophes—agriculture, mining, logging, construction, sewage, and fishing. The conclusion is inescapable: now, even Nature herself demands that salmon be plucked alive from the water and squeezed by a trained technician.

As we watched the video at the Warm Springs visitor center describing the operation of the fish hatchery, my wife Nicole turned to me and asked: "Wouldn't it have been easier to have built a smaller dam or no dam at all?"[1] But such fundamental questions are excluded from the decision-making process before it even begins, for the essence of modern society is the systematic conversion of questions of value into questions of fact. In places like Eastern Europe, where until recently typewriters had to be registered with the police, intellectual freedom is equated with political liberty, but in the United States the disturbing questions are never actually banned by anybody: they are just dismissed as unscientific. For this reason, every step of the dam-building process oozes rationality and logic: engineering surveys and fact-finding committees, public hearings and judicial decisions, regional economic plans and environmental impact statements. The institutions that built the dam and hatchery obtain their revenue from appropriations by elected officials, and their activities are reported on by newspapers and monitored by the courts, so they are answerable to public opinion. Moreover, at any point in the development process, with its endless hearings, committee meetings, and interim plans, public pressure could, in theory, have called a halt or changed the direction of the project, so Warm Springs reflects the perceptions of both the organizations involved and their political constituents. Also, the people who developed and approved this hybrid solution of dam and hatchery think of themselves as rational decision-makers whose policies are based on sound, scientific reasoning, hard-headed economic facts, and streetwise assessments of political realities. Yet, if anything, this massive appeal to reason, science, and democratic institutions is even more disquieting, for if Warm Springs is the product of a rational decision-making process, democratically structured, and supported by the methods of science, then *under what system of cultural assumptions is it rational to destroy a stream with one hand while recreating it artificially with the other?*

Warm Springs, California, in which the destruction of the salmon's spawning grounds is paired with an elaborate technological effort to facilitate its reproduction, epitomizes the contradictions of industrial society. To the environmentalist, industrialism means pollution and habitat destruction, to the businessman it is the creation of wealth, but the Janus-faced image of Warm Springs, with the merging of destructive and constructive processes into a single technocratic icon, is far more representative, for pristine nature and economic productivity are complementary and mutually reinforcing concepts. Just as high-rent districts presuppose slums, and cops require robbers, so the process of domestication and exploitation demands a "raw nature" untouched by the hand of man. It is for this reason that the imagery of the "wilderness" is historically contemporaneous with the development of industry. The worldwide dissemination of industrial technology is presented as voyages of exploration into wild, uncharted regions, both literal and figurative. In the course of the past three centuries, whole continents have been declared "uninhabited," or at best inhabited by *sauvages,* the French word for "untamed"; and even today, in the most industrially advanced countries, millions of tourists a year travel to game reserves and national parks to "experience nature," as if the planetary dynamics of life and death could be confined to a theme park. Thus, in the industrial world view, "nature" is something that exists "outside" of human society, where it is either "discovered" and transformed or set aside and "preserved."

The conquest of nature and the construction of wilderness are two phases of a single industrial process that I call the *one-two punch.* In Phase One, *the conquest of nature*, an area of the earth's surface is first classified as "wild" and "uninhabited," then made "productive" through the application of machines. This phase of the cycle is characterized by stump forests, uranium tailings, foul-smelling pools, and oil-slick beaches. The war cry of Phase One industrialism is "wealth and productivity," and the dominant image is that of the *extraction* of wealth from recalcitrant nature by the application of technique. Moreover, it is the richest, most productive natural habitats that require the most dramatic technological intervention—for it is these that challenge the underlying premise that nature is naturally unproductive and out of control: the buffalo on the Great Plains, the waterfowl flyways of California's Central Valley, or the white-water rivers of the Pacific Northwest. From this perspective, salmon and steelhead trout, which successfully reproduced in California streams by the tens of millions without benefit of biologists or bureaucrats had to be exterminated before they could be saved—just like the American Indian.

Phase One industrialism, with its associated disruption of natural cycles, is by no means a recent development. The widespread use of machines, powered by wind and water, goes back at least to the late

Middle Ages; and, as the historian Jean Gimpel argues, industrial degradation of the environment in Western Europe, including deforestation, water pollution, and smog, was well along by the 13th century. In the year 1255, for example, citizens near the Forest of Wellington were complaining that two limekilns consumed five hundred oaks in one year, while at the end of the 13th century one locality on the Continent banned water-powered saws entirely. By 1230, the British, whose islands were almost completely forested when the Romans invaded more than a thousand years earlier, were importing wood from Scandinavia; and in 1274 "the master carpenter of Norwich Cathedral went to Hamburg to buy timber and boards."[2]

In the mining industry, which is in many respects the prototype of both capitalist methods of finance and mechanized production, people have been aware of the ecological effects of this technology for at least four hundred years. The Renaissance author, Agricola, who wrote one of the first modern books on mining, *De re metallica* ("On Metals"), published in 1556, begins his work with a discussion of the environmental issues. Agricola (c. 1494–1555) was a scholarly German physician who married the daughter of a mining official and spent his life practicing medicine in the mining districts of what is now southeast Germany and northwest Czechoslovakia. The town of St. Joachimsthal, now in Czechoslovakia, where Agricola and his wife first settled, had sprung up less than ten years before but already it was the second most important mining center in Europe after the Tyrol. It must have exhibited the raw vitality one associates with boom towns, for even then, miners were outside the bonds of feudalism, working as paid laborers, often for corporations. Later on Agricola moved to another mining center, Chemnitz,[3] where he amassed a considerable fortune through the shrewd acquisition of shares. In both places Agricola made systematic observations of mining practices, and, significantly, he begins his book with reference to the destruction of forests caused by the production of timber for mine shafts, the poisoning of streams by runoff from tailings, and the effects of metals on the health of the miners themselves—showing that as early as the 16th century educated people were well aware of the environmental costs of the mining industry.

But the destruction of nature is only one aspect of the industrialist program, for once a habitat has been made unfit for human habitation, it signals the Phase Two industrial solution, namely *the construction of analogs of natural processes*, like factories for the breeding of "wild" salmon. In Phase Two industrialism, conservationists set aside the few remaining areas of soon-to-be ravaged landscape as exemplars of pristine nature, say a "wild" river or a game reserve, while in the major portion of the real estate the technocrats ratchet up the system of management and control another notch. That is why the sodbuster and the lumberjack are succeeded by the forester and the environmental engineer. These protagonists are not on different sides, or in different stages of history, but are simply different characters in the same play, succeeding each other as the second act follows the first. Just as in the American West the homesteader preceded the railroad, so in large regions of the Third World today, the landless peasant breaks ground for the World Bank. Phase Two industrialism manufactures the replacements of natural cycles, producing what I call *prosthetic* structures, from a term used in medicine for artifacts that replace lost functions in the body, like eyeglasses and wooden legs. The science and technology sections of newspapers are usually full of one-two punches promulgated as visions of progress. For example, in a recent article on what the authors refer to as the "top 10 coming attractions" in biotechnology, among the innovations listed are genetically engineered crops that contain genes for making pesticides—mutilation and prosthesis folded into one; and genetically engineered bacteria that will clean up oil-spills and chemical dumps by eating the pollutants—a man-made fix for man-made mutilation.[4] In some cases, both phases of the one-two punch are implemented by the same organization. In an advertisement for the Du Pont Corporation, one of the major purveyors of munitions in the Viet Nam War, a Viet Nam veteran plays basketball on artificial legs manufactured by Du Pont.[5] Industrial society is a master of the one-two punch: send nature reeling with a hard Right, then finish it off with the Left.

Industrial society destroys natural cycles with one hand while building fabrications of them with the other, but the integrated operation of these two disparate processes is almost invisible to people because they assume that society is on one side of the equation, nature on the other. In the United States, "raw nature" is conceptualized as unproductive and out of control; and this image is reinforced by endless news footage of earthquakes, hurricanes, floods, tornados, and so on, in which humans are killed or driven from their homes by natural catastrophes, whereas the transformation of sunlight into plants, the generation of oxygen, the fixation of nitrogen, the growth of the body, and the reproduction of the species are usually invisible, background events unworthy of journalistic notice. Even when nature is presented in its more beneficent guise, as in documentaries on the Discovery channel, the imagery simply reinforces the unbridgeable gulf between nature and society. In Sierra Club calendars grizzly bears hunt for salmon in wild Alaskan streams while Bambies cavort in forest glades, but even here the rift between nature and society is replicated in disguised form, for there are no human beings visible at all, as if the earth were still in the Cretaceous.

This industrial image of nature, as either adversary or pristine wilderness, far from being an accurate reflection of the natural world, is a highly selective and culture-bound concept analogous to a style of landscape painting. As in painting, the conventions of the genre determine what is put in and what is left out, what is highlighted, what is attended to, what is defined as foreground and as background. A good example of genre work is the 18th-century painting style epitomized by Stubbs—the stylized scene in which a fastidious aristocrat, in red riding coat and shiny black boots, poses with his pedigreed horse while a pack of eager dogs cavorts in the foreground. Art historians and archaeologists can examine a painting or an ancient artifact and tell you when and where it was made, often within a decade or two; so a cultural style, far from being vague or subjective, is *objective* in the sense that the pictures done in the genre can be measured, photographed, and cataloged, as well as correctly identified by people from other cultures. Yet the genre is a convention nonetheless, for out of all of the things in the world to paint, the culture has decided to include lords and horses and dogs while excluding smokestacks, childbirth, and beggars.

The one-two punch of mutilation and prosthesis, if recognized at all, is usually dismissed as an accidental byproduct of political and economic forces, or embraced as a rational, compromise solution to competing but contradictory demands. Warm Springs dam is the responsibility of a federal government agency, the United States Bureau of Reclamation, which, as its name implies, is charged with reclaiming "unused" water resources for agriculture, which it does by building dams, primarily in the western states; but the hatchery is the responsibility of the California Department of Fish and Game, which is responsible for enforcing fishing and hunting legislation in the state of California, as well as ensuring adequate supplies of game animals through hatcheries, fishing limits, and conservation programs. From this point of view, the one-two punch is the unintended product of contending forces. The farming, mining, and logging industries have enough influence over the government to prevent any serious curtailment of their interests, but the conservationist lobby is strong enough to prevent the salmon from being exterminated entirely. Although this explanation may be an accurate reflection of what goes on in any particular case, it does not address the cultural assumptions which frame the larger context of political negotiation and economic decisions. Why are farming, mining, and logging considered to be "economic" activities in American culture while housekeeping and child-rearing are not? Why are a hundred million salmon only "productive" once they have been canned? Here the conventional wisdom is that the state can tax an industry but it cannot tax non-commercial processes, so it seeks to maximize its power like any other institution. But this answer too is incomplete, for it does not explain why the state chooses to spend its tax money in the essentially uneconomic activity of building and maintaining hatcheries for fish—or why salmon and trout, unique among fish species, are singled out for such special largesse.

However practical a tool or technique may be, it inevitably reflects the cultural assumptions of the people who make and use it, so it always has a built-in symbolic function. In the Warm Springs visitor center, the Bureau of Reclamation presents the dam as a proud example of progress, while the Department of Fish and Game displays the hatchery as an important contribution to the conservation of wildlife. Thus, the Warm Springs dam and hatchery is on one level an example of technology, instrumental to the management of water and salmon, but on another it is a *symbolic* display of the values of the society that created it. As Jacques Ellul pointed out almost half a century ago, industrial society, although often referred to as "technological society," is not defined by its tools and techniques at all—for these are always changing—but by the system of values that organizes the underlying process of technological development.[6] As explicated in the following chapters, the dominant value of contemporary industrial society is in fact the one-two punch itself: the transformation of wild to domestic—and back again—by means of science and "technology."

Warm Springs exemplifies these values perfectly. First of all, the dam is a domestication of nature, conceptualized by friend and foe alike as an obstruction of a "wild" river that once "ran free" to the sea. Second, although it is justified in terms of economics, such as the number of "jobs" that its construction and maintenance "will bring into the local economy," no economic alternatives to the dam-building process are ever really raised, much less weighed in the balance; and this implies that it is the dam itself, not its economic benefits, that is the real agenda. Third, the problem of the destruction of the salmon and steelhead breeding grounds is solved by the application of industrial technology to the breeding of fish, which is itself a process of *de facto* domestication. In short, the adopted "technological" solution displays the dynamic of wild and domestic, untamed and productive, that underlies the entire industrialist cosmology. So, from this point of view, the juxtaposition of dam and hatchery is not a byproduct of the clash of institutions at all but a measure of their profound integration. That is to say, Warm Springs was designed to do exactly what it does: destroying the stream with one hand while recreating it artificially with the other.

The symbolic function of the Warm Springs dam and hatchery is further reinforced by the special role of salmon and trout in American culture. The laws of California classify these species as "gamefish," and the state designates "sport fishermen" as a different legal category from "commercial fishermen." The former must purchase a special sport fishing license to legally pursue their hobby, and they are prohibited from selling the fish they catch. So salmon are legally designated as a "wild fish" in spite of the fact that they are raised in hatcheries and tagged by the Department of Fish and Game. Also, the names of the most highly prized species in California reinforce their symbolic status as wild, noble, and celestial animals: *king* salmon, *silver* salmon, *rainbow* trout, *golden* trout, and *steel* head trout. (*Brown* trout are an apparent exception.) However, there is also a large commercial salmon fishery in California, so the same species of fish exemplify both the sport and commercial categories, thereby demonstrating unequivocally that these categories are creations of culture and not properties of the fish themselves.[7]

Moreover, not every species of fish gets hatcheries built in its honor. In fact, even species with significant commercial value have been fished to extinction, such as the Pacific sardine, which survives now only as brick factories converted to boutiques on Monterey's Cannery Row, so economic considerations are not sufficient to explain the privileged position of trout and salmon. Even in California, hundreds of species are still expected to handle their own reproduction, and many succumb to pollution and habitat destruction unmarked by even an obituary, much less by a boutique. It is the *symbolic* value of salmon and trout, not their economic value, that gives them such a disproportionate share of government resources, for trout fishing in America, long the sport of industrialists and presidents, exemplifies the juxtaposition between the domination of nature and the imagery of wildness that defines industrial culture.

Every year when the rains come in Northern California, the swollen streams wash away the sand bars that accumulate during the summer months, giving the sea-run salmon and trout access to their freshwater spawning grounds. The fish make their way up larger rivers like the Russian and the Eel to tributary creeks and streams, and all along the route are sport fishermen, standing in places elbow to elbow, often waist-deep in ice-cold water, bundled against the morning fog under slate-gray skies, trying to connect with a salmon or steelhead. In campgrounds along the Humboldt and the Noyo, thousands of men gather in their camper vans, usually without their wives, where they light up smelly tobacco that would never be allowed at home, laugh at off-color jokes, and drink more beer than they know is good for them. If asked, they will tell you that salmon and trout are "good eating fish" that "put up a good fight." They will also tell you about the biggest fish they ever caught, about the even bigger ones they've heard of, and the intricate deceptions that they use to catch them. The imagery of this annual migration is men without women catching wild fish that are noble and formidable adversaries. Yet, presumably, the fishermen, many of whom have built the dams and work the pumps, never reflect on the fact that the fish they catch were incubated in stainless steel trays, fed Purina fish chow, and hauled to the sea in tank trucks.

The Anthropological Perspective

The symbolic dimension of industrial activity is ignored by the commonly accepted definitions of economics and technology taught in the schools, which are what anthropologists call *folk theories*: conventional interpretations of events that simply reinforce the unexamined premises of a culture. Because folk theories are elaborations of widely shared cultural beliefs, they have an instant and irresistible appeal to people inside the culture—and are highly resistant to internal criticism—but to the outsider they appear transparently contradictory and self-serving. Almost all conventional definitions of industrial society exemplify in various ways the assumptions of the system that they purport to describe. In most history books, for example, the "industrial revolution" is presented as a "technical development" based on mechanical power and fossil fuels, whereas the inevitable ecological effects of this process, if mentioned at all, are treated as side effects or as "political" issues, as if combustion engines could be discussed independently of the carbon dioxide cycle. In a similar way, to the technocrat and engineer, industrial technology is the application of scientific principles to the development of new materials, processes, and machines—without regard for either the social or natural implications of the technology produced. So too, the economist and financier define industrialism as capital investment in new, more productive technologies—while pointedly excluding effects on the biosphere from calculation of the potential costs. Even when industry is forced to attend to the ecological implications of its actions, as with laws requiring environmental impact reports, these costs are treated as arbitrary externalities imposed by meddling governments, not as "real" factors in the assessment of productivity. Yet it is precisely this cultural tendency to think of technology as completely walled off from the air and water that is one of industrial society's most distinctive and idiosyncratic features. In short, all of the commonly accepted definitions of industrialism replicate the basic premise of the culture—that technology is a self-contained process, answerable only to its own laws, which straddles nature but is not really part of it.

Figure 1.1 illustrates the folk theory of industrial technology, which I call the *technocratic* theory of history. As anthropologists have long recognized, *technology*, defined as the social organization of tools and techniques, is a universal feature of human society, but *technocracy* is a social system in which institutions are organized in reference to the values of scientists and engineers. Although *technocracy* is often used as a descriptive term in comparative government, like gerontocracy or

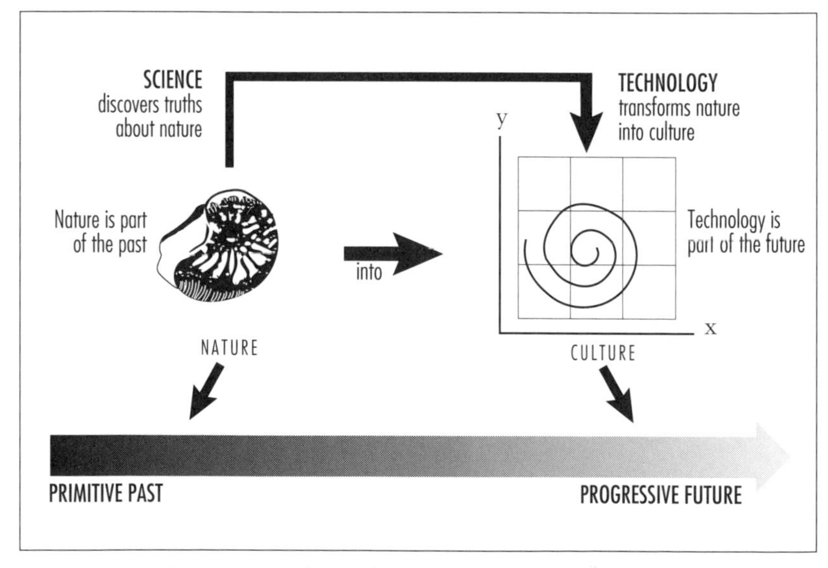

Figure 1.1 The technocratic theory of history.

plutocracy, I use it in a more narrow sense to denote the ideology of modern industrial society, in which social policy and political debate presume scientific models of nature and society, and knowledge itself is reduced to scientific research and description. This technocratic theory of knowledge and society is in turn validated by the technocratic theory of history. In the latter formulation, small-scale societies, such as the Indians that lived at Warm Springs before the valley was flooded, are considered to be representatives of a vanished past, hence closer to nature and earlier in time than ourselves, whereas industrial societies, in contrast, are not only part of the present, that is, "modern," but are on the road to the future as well, which they are actively creating through the reciprocal institutions of science and technology. In the folk model, science is a process of discovery of truths about nature (itself closer to the past), whereas technology is the use of scientific truth to transform nature into culture, creating a richer, happier future.

This imagery of primitive past and progressive future was created by the founders of anthropology in the 18th century, but modern anthropologists have increasingly come to recognize the distortions implicit in this view of history. Although in the popular media anthropologists are portrayed like Indiana Jones, associated with cannibals, savages, and lost civilizations, contemporary practitioners have been much influenced by the fieldwork tradition of social anthropology, which is based on the technique of *participant observation*.[8] In this approach to other cultures, pioneered by Bronislaw Malinowski in the 1920's, anthropologists live for a time in the society they want to learn about. They strike up social relationships with the local people, learn the language, participate in local events, gossip with the neighbors, and generally try to interact as members of the community with the goal of developing a documentary record of the culture based on their own observation and experience. Then at the end of this process, they try to translate the premises of their new culture into terms understandable to people in their own. Since Malinowski's time, there have been hundreds of studies by social anthropologists in all parts of the world; and the view of human technology that emerges from this research is a far cry from the autonomous force of technocrats or the rational efficiency of economists. In the industrial definition, technology is regarded as a branch of physics that stands in marked contrast to such symbolic activities as religion, art, myth, and ritual; but anthropological research shows that technology is systematically related to all of these, such that tools and techniques are embedded in the symbolic system of a culture.[9]

Nuanga: The Integration of Technology and Ritual

A good example of the interdependence of symbolism and technology is the process of turmeric extraction on the island of Tikopia (pronounced tik'-o-pee-a) in the South Pacific, first documented by the anthropologist Raymond Firth. In this society, the technical process of extracting a spice from the turmeric plant can be either a mundane technical activity or a ritual act surrounded by incantations and taboos, depending upon the circumstances under which it is performed.[10] The island of Tikopia is so small that one can easily walk around it in a day, and when it was first studied by Firth in the 1920's it produced all of its own food and materials, being often cut off from the outside world for months at a time. It is hard to find a better example of primitive self-sufficiency, yet the social relations are far more complicated than the materialistic theory of technology would ever have predicted, for fine-grained analysis of the social organization reveals many structural features that are supposedly the prerogative of high technology and civilization.

The Tikopians extract a vermilion pigment from the root of the turmeric plant (*Curcuma longa)* through a process of washing, scraping, sifting, and baking. The technology of extraction is straightforward, although complex and time-consuming, and turmeric pigment is often prepared by individuals enlisting the help of their neighbors and kinfolk. The turmeric produced by these task groups is used as a pigment, spice, or commodity, but it has no special religious or ritual significance. This is in contrast to turmeric produced by chiefs, which is called "the perfume of the gods" and is considered to be endowed with magical powers. This belief is not as exotic as it may first appear but is analogous to the Eucharist in the Catholic Mass, which physically remains bread and wine but has been transubstantiated by the action of the priest, as the anointed representative of God, into the body and blood of Christ. The ritual extraction of turmeric by chiefs is part of an annual ritual cycle, called the "Work of the Gods," and it was observed by Raymond Firth in 1928 and again in 1952.

The turmeric plant is an annual, and it is replanted every year around the same time that it is harvested. The extraction of turmeric from the harvested plants, when it is done by chiefs, is a ritual event called *nuanga*, and it is performed at the beginning of the trade wind season, when the constellation Pleiades rises in the eastern sky just before dawn. One of the chiefs on Tikopia, the Ariki Tafua, told Raymond Firth that the turmeric is ready to dig when the constellation "stands on the shoulder of the mountain" just before dawn, and for this reason the Pleiades are called the "Star of *Nuanga*."[11] Each chief (save one) on Tikopia hosts his own turmeric-extraction party, the *nuanga*, but that of the paramount chief is the most important. The ceremony takes place at certain springs, each associated with a particular chief, that provide the water for the washing. Some men on the island are recognized as especially proficient in the technical skill of extracting the pigment, and one of these acknowledged experts is invited by the chief to participate. Furthermore, this expert is considered to have some special spiritual gifts as well, such as access to a deity whose skill can be called upon in turn. In addition, a number of men agree to process their turmeric at the same time as the chief as part of the common ceremony. To summarize, once a year at a certain time, each chief organizes a turmeric-extraction party, usually by invitation, which meets at his associated spring under the guidance of a ritual expert also selected by the chief.

There is a lot of work that has to be done before a *nuanga* can begin. Aqueducts, made of split areca palm logs, are needed to carry the water from the spring, and these are supported on stakes set in the ground. A grating shed is also erected, made out of thatch, and the ritual expert supervises the manufacture of a filter cloth out of strips of coconut frond sewn together with hibiscus fiber. The expert examines pieces of the cloth as it is being made to ensure that the workmanship is consistent and the mesh is of the proper size. Task groups also go out to harvest the turmeric of the various participants. Afterwards, the roots are washed by a group of women recruited from the families of those participating and formed into two ranks. Seawater is carried up from the beach in bowls by the children.

When all these preparations are complete, the *nuanga*, properly speaking, begins, and the participants are henceforth hedged about with a number of taboos that appear to have little to do with technical considerations. Participants and nonparticipants may not speak with each other unless it is strictly necessary, and they may not sit or eat together. Furthermore, within the *nuanga*, the participants separate by sex, the men staying in the house of the chief, the women in an empty house made available to them. At this time as well, there is a strict prohibition on sexual intercourse, which would ruin the pigment if it occurred, making it soft and mushy; and one job of the ritual expert is to examine the turmeric as it is being made for signs of contamination by semen. Presumably related to this is a ban on soft and mushy foods, like papaya and banana. All food must be cooked in an oven, making it "hot" food—like our own "hot meal" which may be cold by the time one eats it. Food from outside the group is "cold" by definition, and violations of the food proscriptions will cause changes in the turmeric that can be detected by the expert. Revealingly, Firth says he suggested to the people that one spoiled batch of turmeric he observed was due to a failure to add the right amount of oil or to bake the mixture for a sufficient length of time, but no one considered these explanations to be plausible. They insisted that a violation of the taboo against eating "cold" food was the source of the problem.

In the technocratic theory of history, subsistence societies are supposed to have a simple social organization to mirror their simple techniques, but the social organization of the ritualized turmeric-extraction party, the *nuanga*, is far more complicated than the work itself requires. Besides the chief and the expert, it consists of several work groups, each bringing with them several lots of turmeric roots for processing. Each lot in turn is owned by an individual who is a member of the work group. These work groups are voluntary associations of neighbors and close kinfolk, and each work group may include members of different clans (a clan is group of people who believe themselves to be descended from the same remote ancestor). A typical *nuanga* might be composed of three work groups, each composed of four men, each of whom owns a batch of turmeric roots to be processed. However, the work groups are ranked in a hierarchy of priority, with the work group that contains the chief being the highest. The processing of the turmeric of the chief's work group is begun on day one, that of the second group on day two, and so on. Furthermore, within each work group there is a group leader who stands in the same relation to the other members of his group as that of the chief to his followers. On any given day, the turmeric of the leader is begun first, followed by that of the other members of the work group, but the turmeric of different individuals is always kept separate. Furthermore, since it takes longer than a day to process turmeric, the batches of different work groups are in effect interdigitated: thus, on day two, group two is just starting out and group one is beginning its second day, while group three is still waiting. Each of these batches and priorities has a name to distinguish it. The first batch of turmeric produced has the greatest significance, while the last batch is called the "*nuanga* of women," and it is given to them as part of the ordinary domestic supply of paint, used in coloring bark cloth.

The sexual distinction carries throughout the *nuanga*, with some tasks allocated to women, others to men. Men build the shed and aqueduct; women wash the roots. Grating is done primarily by women, but the baking of the final product is an exclusively male domain. Each of these divisions has a task group structure as well, with subgroups of males and females performing specific operations with particular pieces of equipment. For example, at the height of the grating process, when the roots are shaved into thin strips, six bowls and three troughs were in use, as well as sixteen staves and four taro-graters, incorporated into a composite task group of twenty-five women and two men. The operations are also spatially segregated in a complex way that allows certain stages of the production process to be restricted to males. The filtration stage, for example, is off-limits to women. In this operation, the grated turmeric is kneaded with water inside of a funnel constructed out of the filter cloth, which causes the yellow dye to flow out, leaving a fibrous sludge behind. A pole is erected at the entrance to the *tafatafa* enclosure to warn off women. This taboo is only in force while the chief's turmeric is being processed, and women help at other times. The filtration operation is also the occasion for the first major ritual, a prayer uttered by the chief and addressed to the eel god, who has jurisdiction over the fresh water of the springs. The prayer also makes reference to red plants and fish, whose redness can be divinely elicited to enhance the brilliance of the turmeric. Furthermore, it is thought that spirits of the woods will steal the pigment if the protection of a deity is not actively sought. In the *tafatafa* enclosure it is also customary for chiefs and elders to set up objects of great ritual power, like a spear belonging to a familiar spirit, in order to facilitate the technical process.

The action now moves from the *tafatafa* to a sacred enclosure where the filtrate is decanted and the pigment is separated from the spice. The enclosure is dominated by a sacred stone set upright on the ground; and in former times it was said to be off-limits to commoners, leaving to the expert, the chief, and members of the chiefly family the performance of these tasks. Decanting consists of carefully pouring off the water from the filtrate, but the separation operation is trickier. The pigment floats as a dark-red layer of slime on the heavier yellow filtrate. The expert must scoop out the pigment with his fingers, not wasting a drop, while leaving the yellow layer undisturbed.

The red pigment is now mixed with coconut oil and water, then baked in an oven. The pigment is poured into wooden tubes about a foot long and four inches in diameter. These are roughly cylindrical in shape, tapering away slightly from the open end. One end is permanently sealed, but with a small, plugged hole, while the other is left open. In the prayers at this stage, the mixture is equated with the divine excrement, an inversion of its final status when it becomes the "perfume of the gods." After these invocations, recited quietly by the chief, the tubes are then stood on end in an earth oven, lined with hot stones, and the door is sealed with pads of leaves. The mixture is then left to bake for about four hours. When it is judged that the preparation is sufficiently cooked, mats are spread out for the oven deities, and bowls of food are offered to them. Then the tubes are removed and set on the mat. The expert's assistant removes the cooked turmeric by blowing through the hole in the sealed end of the cylinder, thereby forcing a "piston" of hard red pigment out through the top. The cylinders of pigment are then individually wrapped in bark cloth and hung from the beams of the house like salamis.

The Symbolic Perspective

The integration of symbolism and technology, so easily observable in Tikopia, is not just a feature of small-scale societies but was important in the ancient world as well. In ancient Greece, the medical profession was a cult of the god Asclepius, and in China as recently as the 19th century treatises on steel-making discussed the appropriate prayers and ritual garments along with temperatures and ores.[12] But how does industrial society interpret findings such as these? First, it defines small-scale societies such as Tikopia as "primitive" societies, representative of an earlier period of social development, so any observed relationships between technology and symbolism can be dismissed as inapplicable to the modern world. Second, industrial society elaborates a theory of secularization which denies that there is any contemporary connection between myth/ritual on the one hand and technology on the other. In this scenario, technology is sometimes acknowledged to have begun in a religious context of priests and magicians, but then it is said to have *evolved*, through a historical process of secularization, to a demythologized state of pure science and pure technique, answerable only to reason and experiment. Sure, all technocrats agree, the Egyptian pyramids were built by priests, but we are all scientists now; and science and rationalism deny the reality of mythic explanations of human action generally and of technology in particular. Then for good measure, social anthropology itself is debunked as a pseudoscience lacking the intellectual rigor of hard-nosed reductionism. In effect, this *technocratic* theory of technology reinterprets Tikopian culture, imposing its own categories on the turmeric production process to separate, by an extraction process of its own, the pure, physical technique from the muddled irrationality of ritual experts, sacred springs, eel gods, and oven spirits.

There are, however, serious problems with this facile bifurcation of human society into separate spheres of technology and religion. The technocratic theory implies that the social organization of societies with "primitive" technologies must be simple and "primitive" too, but this is far from being the case. In fact, in Tikopia, in the ritual production of turmeric by chiefs (as opposed to the secular production by ordinary citizens), the social organization is far more complex than the physical process requires. Indeed, it exemplifies a number of properties that are associated in industrialism's own theories of technical progress with the emergence of large-scale civilizations: there are task groups composed of lower-ranking members of the community who bring raw materials to a ritual center for processing; convocation of the event by centralized political authority in the form of the chief; and supervision of the technical aspects of the process by a recognized expert who has intellectual skills and specialized knowledge inaccessible to the common run of humanity. In short, when the social structure of turmeric production in this so-called subsistence society is examined in detail, the product is manifestly *not* produced by individual craftsmen working alone, even though this might in theory be possible, but by a complex division of labor involving male versus female groups, aristocrats versus commoners, specialists versus nonspecialists, work groups versus their constituent subgroups—and all of these social distinctions are further subdivided by the concepts of territory, hierarchy, clan membership, and individual ownership.

However, when pressed on the empirical anomalies raised by their own theories of social evolution, the technocrats abandon the field of history entirely and take refuge behind the ramparts of physics. They declare that there *must* be such a thing as autonomous technology, for modern physics asserts that the eel god, the oven deities, and the wood spirits are superfluous to the extraction of turmeric; and if one controlled the temperature properly and measured the amount of oil, it would come out right every time—or at least *most* of the time, with a certain definable margin of error. Moreover, they will claim, the intellectual system of the Tikopians is tautologous, with a flaw in production taken as prima facie evidence that a taboo has been violated, with no attempt to independently ascertain whether it actually was. In short, they resurrect the Victorian idea that Europeans experiment whereas savages are just superstitious. Yet once again, the anthropologist is uneasy. The technology of even nonliterate peoples is based on empirically efficacious techniques that are used precisely because they are observed to work, and many of their technical activities are regarded as no more magical than changing a tire. Even in Tikopia, the manufacture of turmeric by women is largely a secular process with minimal divine intervention. Also, the abstract knowledge that informs industrial production is not as a rule susceptible to empirical testing either, for it is not the workers on the assembly line who determine the truth and falsity of theoretical concepts. The scientific knowledge that is brought to bear in modern high technology may be historically developed by seekers after truth with access to experimental methods, but by the time an idea goes into production on the factory floor, it is accepted as a fact on the word of the expert, just as in Tikopia.

For this reason, I *reject* the technocratic premise that *there is a fundamental difference between technology on the one hand and myth/ritual on the other* and propose instead that the great gulf between these two seemingly disparate phenomena is itself a culturally constructed boundary and not a fact of human sociology. The technocratic theory (Figure 1.1) assumes that symbols are superstructure, determined by underlying physical processes, but symbolic anthropology indicates that myth and ritual are central to human activities and deeply intertwined with even rational modes of thought. Money, for example, from the physicist's point of view, is just a configuration of atoms, yet on a symbolic level it is something far more, instrumental to the rise of empires and the fall of governments, so to define it in strictly technical terms, as pulp and pigment, is to seriously misrepresent its significance. In the same way, bread and wine are physical substances, describable in purely chemical and physical terms, but when incorporated into Christian ritual, they function as symbolic entities that can only be understood in reference to the story of the Last Supper, the theology of the Eucharist, and the gesture of sharing food. Thus, to the anthropologist, a myth is not a falsehood or a fairy tale but rather a story that explains why things are the way they are, defines the kinds of actions that are considered inspirational, and justifies the prevailing norms and institutional structure. As the anthropologist Claude Lévi-Strauss has shown, mythological systems, far from being confused and irrational, are exquisitely logical constructions, exemplifying intellectual processes that are basically similar the world over, irrespective of the so-called stage of cultural development of the societies in which they are found.

So in analyzing contemporary high technology, I have combined the approach of structuralist and symbolic anthropologists, such as Lévi-Strauss, Mary Douglas, and Edmund Leach, with techniques for the analysis of semantics developed by linguistic anthropologists such as Floyd Lounsbury and Harold C. Conklin, extending these methodologies to include nonverbal, visual material.[13] All of these anthropological approaches assume that human symbolism is productive, systematic, and usually unconscious to its practitioners.[14] It is *productive* in the sense that people are constantly applying the symbols of their culture to new situations and combining them in new ways. In language, for example, people form sentences they never heard before and talk about subjects that hitherto had never been discussed. In world languages such as English, hundreds of new words and phrases are added every year to take account of new situations and events: *mutant ninja turtles, S & L bailout, perestroika*. At the same time, this productive use of symbols is *systematic*, for it conforms to cultural norms about how words are to be combined into sentences, and how terms are to be modified to indicate such grammatical distinctions as plurality and gender. Although people use their cultural repertory of symbols in a productive and rule-governed manner, usually they cannot describe either the cultural rules or the circumstances under which they apply. Cultural structures are typically unconscious and invisible to people who have grown up with them—even though they use them perfectly. For example, in the pluralization of nouns in English, native speakers, unless they have had a course in linguistics or in teaching English as a second language, usually tell you to "add s to the end of the word" in order to make it plural; but the rules of the spoken language are a good deal more complicated than this. In English, one adds the sounds *s*, *z*, or *es* to form the plural of regular nouns, as in *cats, dogz*, and *horses*, and the correct choice of the plural suffix depends on the speech sounds that terminate the singular form. To complicate the situation, there are also many irregular nouns in English that either do not mark the plural at all, such as *deer: deer*, or which mark it in a highly idiosyncratic way, such as *ox: oxen* or *mouse: mice*. Thus, the rules that govern the productive use of symbolic systems, although typically unconscious, are anything but simple. Their explication usually requires the efforts of diligent anthropologists with a representative sample of cultural materials. Significantly, the symbolic structures explicated by anthropologists, even when unconscious to participants, are predictive in a scientific sense, for if done correctly they can be generalized beyond the original context, enabling scholars to parse additional samples of cultural materials. For example, a pluralization rule in English should be able to predict the plurals of English nouns that one has never seen before (Exercise 1.1).[15] The generalization of a symbolic analysis to new contexts and materials serves as an empirical check on the validity of the analysis.

Exercise 1.1. Write the rules for pluralizing regular nouns in English.

Using a dictionary, list the singular form of some sample English nouns as a column on the left side of the page and their plural forms on the right. For simplicity, exclude such irregular plurals as *deer* and *oxen*. Then, arrange the nouns into *s*, *z*, and *es* classes, using the phonetic spelling of the words, rather than the standard spelling which can be misleading. Try to predict the plural form from the sound of the final syllable of the singular form. Write down the contingencies in the form of "if...then" rules: "If sound X, then use plural *s*, otherwise..."

To test your hypotheses, add additional nouns to the list to see if their plurals are predicted correctly by the rules.

For extra credit: Write rules for the irregular plurals too. But be prepared to abandon the rules you developed for regular forms—for they may not work in the new context.

Anthropologists have also applied the concept of productive and systematic symbolic structures, similar to those inferred by linguists, to the analysis and description of nonverbal materials. For example, Bernadette Bucher uses Lévi-Strauss's methods to interpret the several hundred copperplate engravings that were published between 1590 and 1634 in de Bry's *Great Voyages*, an epic work on European exploration. By analyzing the contexts of occurrence of graphic elements and the alterations of motifs from one picture to another, she develops a model of the mythology that underlies the art, one that complements but is not identical to the overt content of the companion text. The "linguistic" analysis of visual material is applicable to contemporary societies as well, and, like linguistics, it can be done in the privacy of one's own home with a minimum of equipment (Exercise 1.2).[16]

Exercise 1.2. Parse a visual symbol in *Star Wars*.

Rent the movie *Star Wars* at a video store. With the freeze-frame button on the VCR, stop at each scene where there is a "starburst motif" (i.e., radiant light from a point source, such as an explosion, a sun, or a starfield), and make brief notes on the scenery, plot, and characters that occur with it. When you have done this for the whole movie, review the starburst scenes on the video to be sure you correctly transcribed the situations. Then transpose the descriptions to a large sheet of paper. With a marking pencil, connect all symbols, such as names or characters, that occur in more than one starburst scene. (These symbols are the *level-one associates* of starbursts.) Then review the video to see under what other circumstances these level-one symbols occur, and divide them into classes based on the contexts in which they occur. Try to find symbols that are found only with starbursts but nowhere else, as well as symbols that are never found with starbursts (but do occur with starburst level-one associates). When you have formulated a complete and consistent rule of occurrence of some starburst associate, test your hypothesis using the images in *The Empire Strikes Back*.

The anthropological investigation of symbolic processes cross-culturally, using analytic methods first developed for the comparative study of language, has profound implications for our understanding of technology. Whereas the technocratic theory of technology regards myth as a primitive stage in the evolution of rational thought and ritual as a precursor of technique, symbolic anthropology indicates that the technocratic theory of history is itself a cultural construction, embedded within a larger, semantic structure defined by language, ritual, and mythology.

The Symbolic Body

In all societies, human technology is incorporated into what I call the *symbolic body*. That is, *technical skills and physical objects are coordinated by a symbolic representation of the human body with irreducible mythical and ritual aspects;* and this integrated construction, as shown in Figure 1.2, is then projected back on to the landscape as observable social groupings, artifacts, architecture, and land use patterns typical of a culture. In Tikopia, for example, the process of turmeric extraction is inseparable from such purely social concepts as clans, chiefdoms, and work groups. That is to say, "technology" is not simply a list of techniques, like recipes in a cookbook, but *activities* performed by *people*—and people always have a social organization that they bring to bear on technical tasks, an organization which is represented in Figure 1.2 by the life cycle of social beings. I use the term "social beings" instead of "people" because in many societies animals are included in the definition of social beings, such as house pets in the United

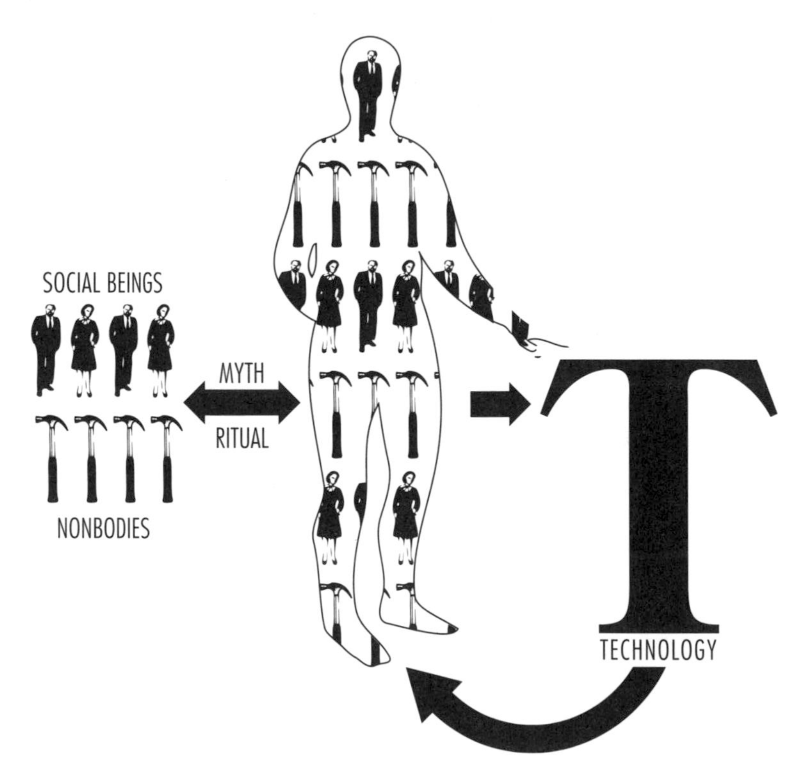

Figure 1.2 Technology is a projection of the symbolic body.

States, whereas human beings are often treated as if they were inanimate objects.

Also, as shown in Figure 1.2, social organization is intimately related to ritual activities. Because human society is largely based on symbolic distinctions that are otherwise unobservable, say the difference between aristocrats and commoners, or between married and unmarried, it is ritual, with its associated rites and emblems, that performs the essential task of creating and marking the various categories of social beings: children, adults, males, females, commoners, aristocrats, married, single, or what have you. In Tikopia, for example, there are rituals that discriminate chiefs from their subjects and aristocrats from commoners, just as in our society there are rituals that create and mark the distinctions between presidents and ordinary citizens.

In addition to the universe of social beings recognized by a society, there is also a sphere of what I call "nonbodies"—namely, those entities that one does not have to deal with socially. A person interacts with social beings by means of social skills such as speech, gesture, and facial expression, but one interacts with the nonbody through purely physical techniques, in which the effect produced is a direct function of the forces applied. This is the domain of technology proper, and in our society, where almost everything has become a commodity, including social processes themselves, almost everything can be classified as "nonbody" and operated on with technical skills. But in many societies, the domain of nonbody is much more restrictively defined, so that even the animals that one hunts may require some ritual or social interaction, such as an apology or prayer, to complement the technical procedure of killing them. Although in all cultures the nonbody typically includes physical objects and raw materials, it is not coextensive with these inanimate objects, for both animal and human bodies are often incorporated into the nonbody framework for such purposes as hygiene, athletics, ritual, and medicine. Thus, the nonbody is the set of culturally defined contexts in which entities, whether animate or inanimate, are operated on using primarily physical techniques. From this anthropological point of view, the *symbolic body* is the culturally specific system of beliefs and symbols that interrelates categories of social beings (produced by ritual) and nonbody entities (produced by technical skills). Although in the technocratic theory of history, social organization and technology are radically distinct categories (or else totally confounded), in my perspective they represent distinct processes, the ritual and the technical, that are complementary, coexistent, and mutually reinforcing.

It is important to bear in mind that the symbolic body as I use the term is not the same as the biological human body, although it physically interacts with it, nor is it merely an aggregation of individual biological bodies, even though individuals participate in it. The symbolic body presupposes individual knowledge and capacity, but it is more than the content of an individual mind. In this respect, the symbolic body is analogous to language, which is also both an individual and a collective phenomenon. We not only acquire our linguistic skills as members of society, but the language itself is far larger than any one individual's knowledge and use of it. An English dictionary contains literally thousands of words that are unknown to any given speaker, and knowledge of a language is transmitted from one person to the next without loss of continuity, usually surviving the death of any particular individual. So language is in one sense an individual phenomenon, in that it presupposes linguistic knowledge in the speaker and the hearer, but in another sense it is a collective phenomenon, distributed among members of a community but not residing in its entirety in any single one of them. The symbolic body, like language, is both an individual and a collective production.

Although the symbolic body presupposes language, it is not exclusively verbal but incorporates what I call *iconic* information as well, such as graphic art and physical objects. For this reason it cannot be unambiguously or exhaustively conveyed in writing or speech but has to be supplemented by analogic mechanisms, such as visual art, gesture, dance, and rhythm. Although the word *icon* is Greek for "image," iconic content is not necessarily visual but can be apprehended through any physiological sense (or combination of them) used to recognize nonverbal forms: touch, taste, audition, vision, proprioception, temperature, deep pain, sharp pain, smell, and whatever. In fact, turning the equation around, one could say that the symbolic body is built up and transmitted by the cumulative effects of such iconically expressed symbolism as dance, graphic art, music, and ritual interacting with a language.

The symbolic body is represented in Figure 1.2 as a human figure into which tools, biological bodies, and the life cycles of social beings are incorporated. The symbolic body organizes social groups, demarcates the stages of the human life cycle, defines the work that is appropriate to each social category, and shows the connections of individual, biological bodies to larger social and cosmic processes. It merges social organization, ritual, mythology, and technique into the integrated construction that is commonly called technology. Although the symbolic body references biological bodies and is shown here as a generic human figure, it differs from human anatomy in critical respects, for it may postulate features with no physical equivalent, such as souls, or deny the significance of physical processes that are essential to human biology. Thus, the symbolic body is a collective representation of the biological body, but it is far more than an aggregate of physical bodies, for it is constructed on the symbolic level of words, images, and ideas. Moreover, it should not be confused with the culture's ideal of the physical body, such as the slim woman and the athletic man, for the ideal physique is yet another manifestation of the symbolic body itself. Nonetheless, the term "body" is appropriate to it because it presupposes symbolic concepts—such as, head and foot, left and right, young and old, skin and bone—*which are only meaningful in the context of human biological bodies interacting with the physical world and with each other.*

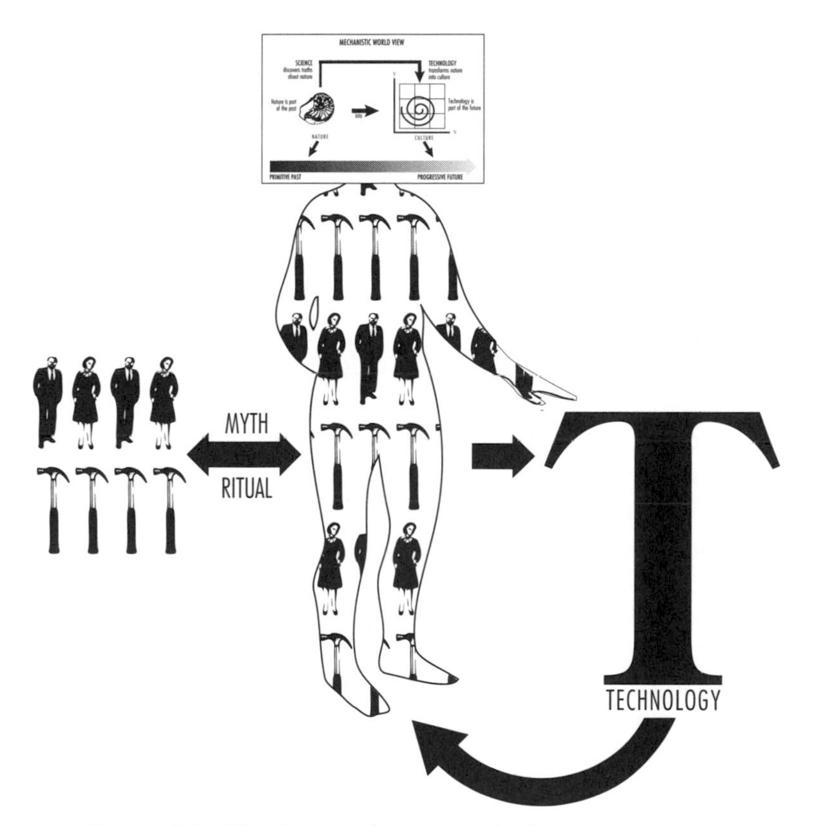

Figure 1.3 The theory of progress is the overt portion of the technocratic symbolic body.

Beyond Technocracy

The concept of the symbolic body shows that much of what Westerners regard as facts of history and objective processes in nature are really just projections of the technocratic mythology itself. As shown in Figure 1.3, the industrial notion of autonomous technology is a folk theory embedded within a social and symbolic framework of ritual and myth. In other words, the technocratic theory of history (Figure 1.1) is not a scientific description of historical events but a mythology that defines the symbolic body of industrial society. From this perspective, the whole technocratic enterprise of science and technology, of mechanism and measurement, of transformations of nature into culture, of primitive past and progressive future—all of these are culturally constructed symbolic categories that interrelate ritual acts to technical skills by means of a mythical charter.[17]

This anthropological perspective can in no way be reconciled with the technocratic theory of history, for the latter maintains that Tikopia (Figure 1.2), with its integration of technology with ritual and myth, represents a primitive stage of historical and technological development. Furthermore, technocrats would argue that the progressive transformation of nature into culture by means of tools and techniques (Figure 1.1) is a true description of evolutionary processes and not simply a projection of industrial ideology. Moreover, we are assured that in the temples of the technocracy itself, in the corporations and universities, in the institutes and government agencies, myth and ritual have long been abolished, and the practitioners are now engaged in purely rational pursuits grounded in empirical evidence and scientific theory. Also, this contemporary technocratic system is open-minded and progressive, such that new ideas are constantly being introduced, critically evaluated, and adopted when proven scientifically.

Yet the analytic model of Figure 1.3 leads us to the opposite conclusion, that the concept of rationality itself, as well as its infrastructure of science and technology, are all part of a culturally constructed symbolic framework that blends both factual and mythological elements. Moreover, it indicates that the mythical charter and the ritual acts are still alive and well, not just on Sunday mornings on the other side of town, but enshrined at the very heart of the technocratic system where an anthropologist would expect to find them. Moreover, far from being open-minded and empirical, the technocracy, as would be expected from Lévi-Strauss's theory of myth, is a closed intellectual system, a tautology. Admittedly, it is a huge, imposing tautology but a tautology nonetheless.

The anthropological perspective also leads us to question the stated objectives of the technocratic enterprise. In the technocratic theory, technology takes control of natural cycles in order to create a better future, taming nature and bending it to our will, so that the output of the industrial system is conceptualized as progress itself. But the symbolic body includes both the technological *con*-struction of the future as well as the technological *de*-struction of "natural" and "primitive" cultures and habitats, so the projection of the system necessarily contains both processes as well. In other words, the empirically observable output of industrialism is not progress but the one-two punch. This underlying dynamic is shown in Figure 1.4, where (1) the destruction of nature and the denial of natural cycles create (2) a need for prosthetic devices that simulate nature and which (3) create an illusion of forward motion that (4) rationalizes and confirms (1) the destruction of nature and the denial of natural cycles... and so on. Thus, "progress" is just one more term in the folk system, a purely symbolic category with no more objective reference than the wood spirits and oven deities of Tikopia. The contemporary industrial system is logical, comprehensive, self-sustaining, and self-validating—provided that one never questions the underlying assumptions.

It is obvious that this anthropological interpretation of technology is completely irreconcilable with the technocracy's theory of itself, so we are faced with a dilemma: Do we take the conventional approach and

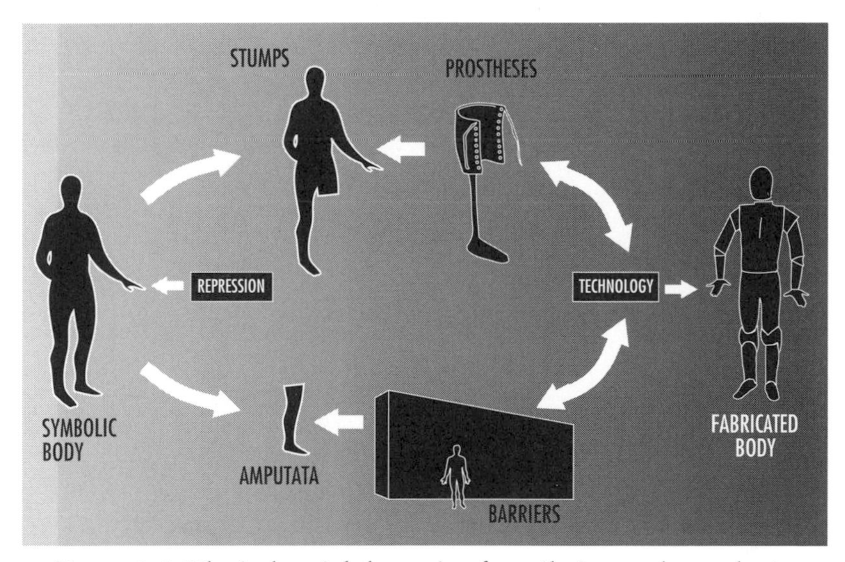

Figure 1.4 The industrial dynamic of mutilation and prosthesis.

continue to bury the discrepancies in the basements of museums, or do we pursue the more demanding course of critically examining the intellectual premises of industrial society? Once again, the system has weighted the dice, for to question the intellectual premises of industrialism, one must be a certified intellectual, and to be a certified intellectual one must spend the better portion of one's life applying parodies of the scientific method to phenomena to which they do not apply. If anthropologists played by these rules, they would still be lining up natives in the district commissioner's office and measuring their heads with calipers. So instead of bestowing a posthumous dignity on the premises of industrial society by engaging it in debate on its own terms, we make the simplifying assumption that the technocratic model of nature and society taught in the universities is a folk system of belief, with no more claim to universal validity than any other theory created by savages. In other words, the question is not whether anthropology measures up to the intellectual and scientific standards of the modern academic tradition but the extent to which the scientific tradition itself can be explicated within the theoretical framework of field anthropology. In short, we put the computer and the spaceship on the same plane as Navaho pots and Zulu spears, and subject industrial culture to the same sort of objective scrutiny that it so routinely imposes upon others. Let us track some mud through the scientific temple, pry the rubies from the idols' eyes, weigh the priests, and boil their bones. Then, to top off our collection, let us stuff a brace of modern intellectuals, dress them in their native garb, and mount them in the lobby.

NEBUCHADNEZZAR'S DREAM

The Mythology of Technical Progress

The intimate relationship between technology on the one hand and religion on the other, which anthropologists have observed in non-Western cultures, is also an essential part of modern industrial society. In the 17th century, Francis Bacon, lord chancellor of England and an influential man of letters, advocated government support for a massive research and development effort to increase the human species' dominion over nature. In a posthumous work of fiction called *New Atlantis*, he sketched the design for a society run by professional organizations of scientists and technicians dedicated to "the knowledge of Causes, and secret motions of things; and the enlargement of the bounds of Human Empire, to the effecting of all things possible."[1] He saw this effort as the Great Instauration (an archaic term for "renewal") foretold by the prophet Daniel, so that in Bacon's mind technical progress was the fulfillment of divine prophecy. Thus, in the original formulation, industrialism was a moral and civilizing force; and in the English-speaking countries in particular, the tales of scientific heroes resonated with an optimistic and expansionist Protestantism that gave technical innovation the force of religion.[2]

Coincident with these developments, facts of history and geography were assimilated to the stages of history presented in the Book of Daniel.[3] As recounted in the Bible, King Nebuchadnezzar of Babylon, in the second year of his reign (604 B.C.), had terrible dreams, "and his mind was so troubled that he could not sleep."[4] As he lay on his bed, he saw a huge image, bright and dazzling, that towered before him: the head of the image was of fine gold, its breast and arms of silver, its belly and thighs of bronze, its legs of iron, its feet part iron and part clay. As the king watched, a stone was hewn from a mountain, not by human hands, and it struck the feet of the image, shattering them, and breaking the figure into fragments until not a trace remained. Then the stone grew into a great mountain that filled the whole earth. The Hebrew prophet Daniel, who had been carried off to Babylon as a boy by the conquering army and sent to Chaldean schools, interpreted this dream for the king, explaining that the head of gold was Nebuchadnezzar himself (Daniel was a professional diplomat) and that the other metals were the kingdoms that would follow his reign. Each would succeed the previous one until God establishes a sixth kingdom that will endure forever, bringing the succession of empires to an end.

The Book of Daniel was written several hundred years after the events it describes, during the Greek administration of Palestine when it was ruled by successors of Alexander the Great; and this image of a descending scale of metals representing the eras of human history can be found much earlier in Greek mythology.[5] The poet Hesiod (8th century B.C.) wrote about a remote golden age more akin to the biblical Eden, with people leading a more natural and contented life. In Hesiod's chronology, the golden age degenerated into an age of silver, in which suffering was more rife, followed by a warlike but chivalrous age of bronze. According to Hesiod, the bronze age was followed by a heroic age of the ancestral Greeks of which Homer sang, finally terminating in the lackluster age of iron, the period in which the author was writing. There were permutations of this model too. The Latin poet Ovid, around the end of the 1st century A.D., describes a historical sequence of raw materials but leaves out any mention of a heroic period, while five thousand miles away, a Chinese sage writing somewhat later, in 52 A.D., inserts an age of jade between stone and bronze. Thus, the biblical story of Nebuchadnezzar's dream may well represent a Hebrew transformation of a prominent Greek chronology, one that replaces a pessimistic cycle of degeneration with a reaffirmation of Jewish faith in the future kingdom of God.

This image of history as a succession of technical stages, which first appears in documented form in the 8th century B.C., was transformed yet again in the 18th century A.D. into a secular theory of social change that is with us to this day.[6] The founding fathers of modern social science were well-educated men, social philosophers and antiquarians, who were conversant with the Bible and who had read extensively the works of Greek and Latin authors. Both Daniel and Hesiod would have been familiar to them, and in developing the concept of historical change, they incorporated the ancient succession of ages, each represented by a characteristic metal, into a new theory of material progress that was scientifically supported by the first systematic study of the past. In the peat bogs of Denmark, the tannic acid in the soil preserved the bodies of ancient agriculturalists so perfectly that some of the first finds were investigated by the police as homicides. In 1807, inspired by these excavations, which yielded not only mummies but clothing and tools as well, the Danish government set up a Royal Committee for the Preservation and Collection of National Antiquities. By 1819, a Danish National Museum opened its doors to the public, and significantly, its collection was arranged into a tripartite system of Stone Age, Bronze Age, and Iron Age, succeeding each other in chronological order—an arrangement, it should be noted, which preceded any scientific method for dating the prehistoric sites. Although the scholars who created this system were well aware of its discrepancies, writing as early as 1813 that "without any doubt the use of stone implements continued among the more impoverished groups after the introduction of copper, and similarly objects of copper were used after the introduction of iron...," they nonetheless represented the historical changes they saw as a succession of ages, not gold-silver-bronze-iron-stone, a cycle of degeneration, as the ancients had done, but as a positive permutation of the classical series—a progressivist vision of technology advancing in strength and power from stone to bronze to iron.

As shown in Figure 2.1, this image of the male human body demarcated into stages is the primary image in the symbolic body of industrial society, and it is this image that defines the relations between nations, establishes the leadership of one industry relative to another, and justifies the development of new technologies regardless of their social and ecological effects. These two images—noble head and forward motion—are definitive of industrial progress.

The three-age theory of history, where periods are represented as a succession of materials, each denoting a stage of social development, was subdivided in the 19th century into a complex stratigraphy subsuming all aspects of human culture. In this still-popular approach to history, people like the Tasmanians or the San (Bushmen) of South Africa, who live nomadic lives as hunters and gatherers, are said to represent the savage stage of human history. Larger and more warlike nonliterate societies, like the pastoralists of the Asian steppe or the warrior states that overran the Roman Empire, exemplify the barbarian stage of history, while those societies with writing and a centralized government represent the highest level of human evolution, that of civilization. The American anthropologist Lewis Henry Morgan, who is still highly respected in the profession for his pioneering studies of kinship, developed

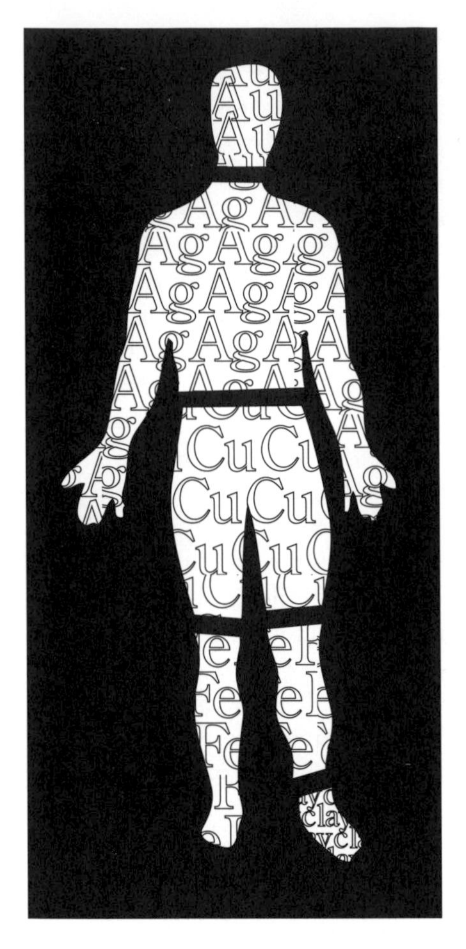

Figure 2.1 In Nebuchadnezzar's Dream, stages of history are demarcated by segments of a human body, each represented by a different artifact or raw material.

this idea into a theory of progressive social evolution defined in terms of technical innovations. In Morgan's evolutionary scheme, first published in 1877,[7] the three stages of Victorian anthropology, namely savagery, barbarism, and civilization, are further subdivided into lower, middle, and upper, and each is accorded a distinct technical achievement:

- Lower Savagery, from the emergence of the human species to the discovery of fire
- Middle Savagery, from the discovery of fire to the invention of the bow and arrow
- Upper Savagery, from the bow and arrow to the invention of pottery
- Lower Barbarism, from pottery to the domestication of animals
- Middle Barbarism, from the domestication of animals to the smelting of iron ore
- Upper Barbarism, from the smelting of iron to the development of the alphabet
- Civilization, from writing onward.

Morgan's social and technological epochs incorporate very disparate events that cannot be treated as points on a single graph. Fire is the human control of a physical state change in nature; the bow and arrow is a borderline machine; domestication is the incorporation of animals into human social groups; and writing is an intellectual skill. Each of these innovations can arise independently of each other—and they probably did. Significantly, these and other "universal" stages of technological development, presented in older textbooks such as V. Gordon Childe's once popular prewar works, *Man Makes Himself* (1936) and *What Happened in History* (1942), are based almost exclusively on the cultures of western Eurasia. The archaeologist Glyn Daniel reports that "I once tackled Childe about his neglect of the civilizations of Nuclear America, and he dismissed my question with the words, 'Never been there—peripheral and highly suspect.'"[8] Yet subsequent archaeology has demoted the three-age theory of history (Stone Age, Bronze Age, and Iron Age) to a local chronology of convenience for the Mediterranean area, and the universal stages of metallurgical development have diversified into a host of regional traditions and techniques.

In fact, what was claimed to be objective scientific research a century ago now appears transparently racist.[9] Morgan, for example, did his major work while the United States government was engaged in the final extermination of American Indian cultures, and his strong theoretical emphasis on the smelting of iron probably has more to do with the politics of colonialism than with his interest in technology. Iron smelting never developed in the New World, and copper metallurgy was a late invention, arriving in Meso-America from Peru around 900 A.D., after the great cities of the Classic period had already gone into decline. The New World societies, independently of the Old World, had invented the lost-wax method of casting and used it to create gold jewelry of spectacular quality, but the cities and temples had been built with stone tools. Yet by Morgan's criteria, the urban societies of the New World could at best represent the stage of Middle Barbarism. Significantly, Morgan debunked the idea that American Indians were capable of creating a civilization on their own. From his observation of Iroquois Indians living on reservations in New York State, he concluded that historical accounts of the Spanish conquest of Mexico were gross exaggerations. Tenochtitlán, he argued, was not a city of paved streets, stone buildings, causeways, and canals but a very big village; Montezuma was not a king but a tribal headman; and the grand banquet described by Cortez was a communal cooking pot. Extensive archaeology has subsequently demonstrated without a doubt that the Aztec capital was a city with a population of 100,000 or more at the time of its demise—five times larger than the London of the day—and its ruler a king, but the best riposte to theories of this kind is the remark of a contemporary historian who commented on Morgan's book: "Why should Mr. Morgan, who never dined with Montezuma, know so much more about such things than Cortés and Bernal Diáz, who did."[10]

The image of the past as a progressive sequence of stages based on technical innovations is a modern version of an ancient myth, what I call *the myth of Nebuchadnezzar's Dream*, and every one of its propositions is contradicted by a century of archaeology, field anthropology, and historical scholarship:

- Even metallurgy, to which the stage theory was first applied, is better explained as diverse pathways of development than as a sequence of stages.
- The evolution of technology is not an irreversible process, for the archaeological record indicates that human societies can become less complex over time as measured by the range of artifacts.
- There is no relation between the physical materials used by a society and other measures of social complexity.
- "Civilizations," defined archaeologically as literate societies with monumental architecture and a class of full-time non-food producers, do not necessarily have more complex tools and techniques than their uncivilized neighbors.
- Many innovative, technically advanced techniques have been developed by small groups of traditional craftsmen using hand-tools.
- Handmade tools and weapons, far from being trivial inventions, are significant sources of power that have played a major role in political history.
- The invention of machines does not in itself usher in an "industrial revolution" or transform the social organization.

Pathways Not Stages

Even when technology appears to unfold as a sequence of stages, the sequence may vary from one place to another, vitiating the concept of an inevitable sequence.[11] Consider, for example, the development of steelmaking in China and the Mediterranean. The Mediterranean method of smelting iron from ore produced a low-carbon metal, called bloomery iron, that was soft and ductile but had to be hardened through hammering.

In order to produce steel, carbon was added to the bloomery iron, a process called carburization. The chemistry was not understood, so carbon was not recognized as the critical element, but blacksmiths knew that the bloom had to be "purified in fire," as they expressed it, in order to make steel. In China, however, the large-scale production of iron was developed through the invention of the blast furnace around 600 to 400 B.C., which forced a blast of air over the molten ore and yielded a high-carbon metal that was just the opposite of European iron. Where the bloomery iron of Europe was too soft because of its low carbon content, the cast iron of China was too brittle because of its high carbon content. Where the metallurgists of the West added carbon to the iron metal in order to produce steel, the Chinese developed techniques for removing it. Moreover, the Chinese developed two different ways of decarburizing iron, as well as a way of combining high-carbon and low-carbon iron in a forge to produce a composite metal with the virtues of both. Thus, the "Iron Age" in the Old World was achieved by different techniques in different places, by bloomery iron in the Mediterranean and by cast iron in China; and these are only the most historically significant pathways. Archaeological investigations show that iron metallurgy in Thailand antedates that of China, and may have inspired it, while in Africa south of the Sahara indigenous iron-making techniques, involving departures from both Mediterranean and Southeast Asian models, were under way by 400 B.C. Thus, the age concept is misleading even when applied to a single material because it ignores the differences in technique that are employed in different places to achieve a similar end.

If the histories of different metals are compared, the prevalence of distinct developmental pathways becomes even more striking. In Europe and the Near East, metallurgy, for the first few thousand years of its history, is copper metallurgy, with only the incidental production of iron as a byproduct of copper smelting, and the precious metals, silver and gold, do not come into prominence until the rise of the "Bronze Age" states of Egypt and Mesopotamia in the 3rd millennium B.C. But in South America, on the west slope of the Andes where the Inca empire later arose, metallurgists were interested in gold from the very beginning, and they developed all manner of techniques for smelting, alloying, soldering, casting, and plating it. Furthermore, in South America, the sequence of metallurgical development is inverted or compressed compared to the Mediterranean. Within the span of a few centuries, alloying, soldering, fusion welding, heat welding, annealing, and the smelting of metal from ore all appear in South America, but the metallurgical traditions are nonetheless fundamentally different from the similar developments that had taken place thousands of years earlier in the Mediterranean area. Although both of the important Old World bronzes, copper-arsenic and copper-tin, were independently discovered in South America, leading to a proliferation of metal tools made by both forging and casting techniques, again just as in the Old World, the historical sequence is exactly the reverse. Whereas Old World metallurgy basically began with copper, followed by bronze, and the precious metals were added later to the repertory, in South America the invention of bronze grew out of an already sophisticated technology for handling the alloys of silver and gold. Alloys, in fact, developed in the Andes as solders, not as a way of changing the mechanical properties of the metal, and the Andean culture, unlike the Eurasian, emphasized assembly rather than casting, constructing votive figurines through the use of sheet-metal techniques. So metallurgy, far from proving the three-age theory of history, demonstrates its inadequacy for even a single material.

Technology Is Not Irreversible

In textbooks from only a few decades ago, the types of materials used to make tools map directly on to stages in the development of social organization: Paleolithic Savagery, Neolithic Barbarism, the Higher Barbarism of the Copper Age, and Early Bronze Age Civilization. The essential idea of social evolutionary theories is that technology and society necessarily move from the "simple" to the "complex," and therefore the "simplest" societies existing today provide the greatest insight into the distant past. But the "primitive hunter and gatherer" fares no better at the hands of archaeologists than the three-age history of metals. Captain Cook, when he visited the large island of Tasmania off the southern coast of Australia, in the late 1700's, judged its inhabitants as the most "primitive" people he had seen in all of his extensive voyaging, describing how they lived in small groups or isolated families with few artifacts but sticks and stone tools.[12] This discovery seemed to confirm the theory of progressive evolution through technical change that was just then being promulgated in educated circles in Europe. Although the Tasmanians were rapidly exterminated on the main island at the beginning of the 19th century, surviving only on the smaller islands of Bass Strait, subsequent scholars saw them as a remnant of the earliest stage of human society, preserved by isolation in what to Europeans was the most remote corner of our sphere. Yet extensive archaeological excavation over the past few decades reveals that Tasmanian technology steadily simplified over a 12,000-year period, as measured by the range and diversity of tool types.

Tasmania, an island about the size of Sri Lanka, lies two hundred miles southeast of the state of South Australia. During the last ice age it was connected to Australia by a broad plain, but as the glacial ice melted and world sea level rose, Tasmania became an island, cut off from the mainland by the waters of Bass Strait. The sea proved an effective barrier to influences from elsewhere. Several species of animals that became extinct on the Australian mainland survived on Tasmania, and the wild dog, the dingo, which colonized the continent about 5,000 years ago, never made it to the island. In the middle 1960's, Australian archaeologists began to investigate ancient habitation sites on the island of Tasmania, using modern techniques of radiochemical dating. At Rocky Cape on the northwest coast are two caves, cut in the cliff face by wave action, that were lived in thousands of years ago. Together, the two caves span more than an 8,000-year period of continuous human use, and they convincingly counter the tenacious idea that isolated people with simple technologies represent the ancestral condition.

When discovered by Europeans, in the 18th century, the Tasmanians had the simplest tool kit of any contemporary people known to anthropology. Their entire material culture consisted of wooden spears with fire-hardened tips, digging sticks, grass baskets, water containers made from kelp, cloaks of kangaroo skin, shell necklaces, bark boats, wooden spatulas, fire sticks, and stone tools. Even this tool kit is far more complicated than it first appears to be, but the important point is that the Tasmanian technology of historical times is a significant simplification from that of 8,000 years ago. The Rocky Cape caves show a much wider range of stone tool types than occurred in historical times, such as a half dozen differently shaped scraping tools, used for woodcarving. Moreover, there was a diverse bone tool kit at Rocky Cape, including punches, awls, needles, and scrapers in a variety of sizes and shapes. Yet over the period from 6000 B.C. to 1500 B.C., the bone tools declined in use until they disappeared entirely from the culture. Also, the Tasmanians stopped eating fish about this time, as confirmed from other archaeological sites, and when discovered by Europeans two centuries ago they disclaimed all knowledge of how to catch fish or prepare them, even though the same species, whose bones occur by the thousands in the lower strata of the caves, survive to this day in the ocean at the foot of the cliffs. One archaeologist has suggested that the disappearance of fishing and bone tools are related to each other, hypothesizing that the bone needles and awls were used to make gill nets;[13] but whatever the reason, the Tasmanian tool kit was less diverse in historical times than it was 8,000 years ago, which indicates that it was already very different from the technology of mainland Australia, which predates it: there are no boomerangs, no spears with fitted stone heads, no axes with wooden handles, no barbed spears, no sewn garments of possum skin—all of which occur elsewhere in Australia at that time. Since the ancestors of the Tasmanians presumably walked to the island from South Australia, they probably arrived with the technology of the mainland people but lost many of these arts and crafts during the subsequent period of isolation. Thus, the Tasmanians are not an example of primitive simplicity, preserved in a remote part of the world, but the culmination of 10,000 years of niche specialization.

The same type of cultural change is also evident among the successors of the Upper Paleolithic peoples in France.[14] This period was characterized by great cultural innovation and artistic achievement, epitomized by the cave paintings of Lascaux; but it gave way, without invasion by foreigners, to a lackluster era called the Azilian, named after the town of Mas d'Azil, where the remains were first excavated. As with the Tasmanians, the bone tool industry was reduced to a handful of tool types, and the flint artifacts deteriorated in workmanship. Although the people continued to occupy the same caves, the tradition of mural art disappeared completely and the portable art evolved into a few stylized geometrical designs. In the cultures that followed them, the artistic tradition seems to have disappeared entirely.

Contemporary hunters and gatherers are probably no more representative of the peoples of Ice Age times than are the Tasmanians. With the exception of Australian Aborigines, who occupied the whole continent of Australia before the British colonization, hunting and gathering peoples are generally tiny populations living in marginal habitats, and they are by no means typical of even preindustrial societies. The people with the best claim to being representative of Ice Age societies, the Australian Aborigines, until recently living a hunting way of life with little outside interference, do not form "simple" nomadic bands at all but have the most complex kin-based social systems known to anthropology.[15]

In the light of the archaeological record in Australia, as well as from historical sources describing other cultures, the technological simplicity of "primitive hunters and gatherers," as they existed until recently, was necessitated by their nomadic life style and made possible by their geographical isolation, often in severe habitats, and it is not an inheritance from prehistoric times. Many of these peoples are probably voluntary dropouts from more constraining societies or remnants of larger populations driven into inaccessible regions by militaristic neighbors. In fact, the "isolated hunter and gatherer," epitomized by the San (Bushmen) of the Kalahari Desert, is a sociological fiction because all of these societies are colonized peoples, living at the grace of industrial states; and none of them have ever been observed under isolated or pristine conditions, if indeed such conditions ever existed. Contemporary hunting societies are invaluable for providing insights into the ecology of hunters and for illustrating how tools and techniques, like spears and snares, might have been used in Ice Age times, but they are very poor models of prehistoric economy and social organization.

The Imagery of Stone and Steel

In the progressive theory of human evolution, there is a hierarchy of physical materials, with stone on the bottom and steel on the top, but this stratification reflects the politics of colonialism, not the laws of physics. Steel was the high-tech material during the flood tide of European imperialism, and the differences between stone and steel became symbolic of the social relations between industrial powers and what is known as the Third World. Like all good myths, it resonates to self-evident truths, and from the 18th century on, every confrontation between Whites and Blacks seemed to hammer home its evolutionary thesis.

For example, metallurgy never spread to Australia and New Guinea until the end of the 18th century, and Europeans did not even know that the highlands of New Guinea were inhabited until the 1920's, when they happened to fly over them and saw that the thick rain forests of the coastal mountains gave way in the interior of the island to grassy valleys and high plateaus densely populated by farmers.[16] In 1933 two Australian gold prospectors landed a light aircraft in a field at what is now the town of Mount Hagen, bringing with them a variety of valuables to establish good relations with the natives, including large cowrie shells, a traditional form of currency in the area, and steel axes, which no Highlander had ever seen before.

Ongka, a local resident whose autobiography was later tape-recorded and translated from the Melpa language by the anthropologist Andrew Strathern, was a young boy when the first airplane passed overhead. "A thunderclap gone mad" one man described it, and Ongka remembers clearly that he was down at the stream with several old men and boys grinding the blades for stone axes.[17] In Ongka's part of the Highlands, not even the stone ax-blades were available locally but had to be traded from people living in the Jimmi River Valley, giving them women and pigs in exchange, and the stone had to be laboriously ground down into the shape of an ax. Even then, Ongka complains, the tool was not very satisfactory for clearing a tropical forest: "Only strong men could finish the task of cutting a tree down; weak men had to give up."[18] And when Ongka's father traded one of their best pigs for a cowrie shell brought by the white man, Ongka was furious: "Why did you get a worthless shell? With a steel axe think of the work we could do."[19]

The rapidity with which iron or steel replaced stone, both in colonial times and in the ancient world, has been interpreted in terms of the great difference in technical effectiveness between the two materials, but experimental studies using stone tools have forced anthropologists to question even this so-obvious truth.[20] The stone axes of the European Neolithic, for example, have been found to be very effective in clearing European forests for farms; and obsidian arrowheads have been found to be *more* effective than steel in penetrating flesh and hide. Also, an anthropological field study in the New Guinea Highlands concludes that the steel ax, far from being dramatically superior to stone in technical effectiveness, is rather more efficient in an economic sense because it has a longer useful life than stone, especially when sharpened with a local grindstone rather than with a steel file, and it needs to be sharpened less frequently. Therefore, its replacement cost is less when measured by the number of hours of plantation labor needed to procure it, as compared to the number of hours needed to produce and maintain the traditional implement. A fine-grained study of the use of stone axes by Amazon Indians came to a similar conclusion: ground stone axes can chop down a tree perfectly well, especially if combined with the use of fire, but they do need a lot of resharpening.

Engineering judgements are always made within the context of economic and political considerations, and even Ongka's testimonials need to be taken with a grain of salt. Perhaps Ongka reasoned that if he took the white man's axes, he would no longer have to give women and pigs to the people of the Jimmi River Valley? Perhaps Ongka, who went on to become his district's first elected representative in the national parliament, is astute enough politically to know that white men love to hear about the power of their tools?

These sociological and political factors make the archaeological boundary lines between "old" and "new" technologies, which are so transparently obvious to the progressive evolutionist, very difficult to interpret. For example, older works on archaeology tell us that East Asia, all through the Pleistocene, was very slow in adopting improvements in flint technology pioneered in west Eurasia, maintaining "obsolete" tool-working traditions for tens of thousands of years. Assuming it is true, does this reflect cultural backwardness or its opposite—the presence of more efficient social and economic systems for putting finished tools or supplies of raw flint into the hands of the consumer, such that the improvements in technique were perceived as of only marginal value?

If technical explanations are so tenuous in the confrontation between stone and steel, social theory must be doubly cautious in attributing major economic and social changes to improvements in tools and techniques. In the traditional theory of technological evolution, for example, the development of the city state is attributed to the rise of agriculture, as if one could cause the other; and V. Gordon Childe, one of the most forceful expositors of this approach, has tried to provide a "technical" definition of civilization in the ancient Near East to substantiate the stage theory of history. His list of "technical innovations" is very revealing: the plough, the wheeled cart, traction animals, the sailing boat, the smelting of copper, and irrigation—technology to be sure, but almost all of these inventions are now known to antedate urban societies in the region by long periods of time, sometimes by thousands of years, and all of them can be found in folk technologies.[21] He has to add the solar calendar, writing, processes of reckoning, standards of measurement, specialized craftsmen, city life, and finally, his sine qua non—surplus foodstuffs supporting a leisure class of full-time non-food-producers. Even by Childe's "technological" criteria, the distinctive features of "civilization" are intellectual skills and social organization.

The Hand-Tool Civilizations

The word "civilization" is still used by archaeologists as a technical term to denote a socially stratified society with full-time craftsmen, monumental architecture, and a literate ruling class; but there is no relationship between civilizations socially defined and the kinds of materials that are made into tools.[22] In fact, when the technical skills are considered simply as techniques, leaving aside the political systems that coordinate them, then there is not even any intrinsic relationship between so-called civilized social organization and the complexity of the technology at all. Even the literate and urban societies of the ancient world were as dependent upon hand tools and muscle power as any New Guinean villager. Prior to Hellenistic times, civilizations lacked machinery and machine tools, and their repertory of technical processes was comparable to the subsistence skills of village communities in Asia, Africa, and the Pacific today. In ancient Egyptian civilization, for example, which conventionally begins in 3100 B.C., there is very little difference in technical complexity between the task groups and techniques depicted on the murals of the temples and tombs and those that can be seen today among village farmers in the tropical Pacific:[23]

- Hunting waterfowl from a reed skiff with a throwing stick and live decoys;
- Drying and salting meat for storage in earthenware jars;
- Fattening wild-caught cranes on pellets of cooked dough;
- Weaving linen from flax fibers on a horizontal loom;
- Producing paper by pounding on the pith of the papyrus reed;
- Grinding grain by using a stone roller and stone anvil;
- Plowing a field with a wooden plow drawn by a pair of bullocks;
- Melting gold in clay crucibles;
- Building a boat by binding together bundles of buoyant reeds;
- Brewing beer in pottery tubs;
- Trussing up cattle with ropes preparatory to butchering them;

...and so on.

Although Egyptian crafts were coordinated by a literate bureaucracy under the auspices of temple and king, the skills themselves are well within the competence of small, face-to-face groups of people recruited from a local community. Even something as fundamental as the grinding of grain was no more "complex" in ancient civilizations than it was among their nonliterate neighbors. In the temple cities of both Egypt and Meso-America, grain was hand-ground with a quern, by women arduously rotating one stone on another.

Nor is there any evidence that the classical civilizations of the Ancient Near East possessed machines, which seem to have been first developed around the 4th century B.C.[24] In ancient Egypt and Mesopotamia, the closest thing to a machine was a hand-powered device for transferring a bucket of water from a canal to an irrigation ditch by means of a long pole. This device, called a *shadoof*, can still be seen in the rural areas of India and the Near East, and it consists of a watertight vessel on the end of a rocker beam. The operator lowers the bucket by raising the opposite end of the beam, dips it into the canal, raises it up again to bring the bucket into position over the ditch, and then empties it by lowering the beam once again. A device that has been interpreted as a *shadoof* is depicted on a Mesopotamian cylinder seal of 2500 B.C., near the very beginning of Near Eastern civilization, but the *shadoof* is nonetheless closer to the bow and arrow than to a true machine. Although it uses mechanical principles, it exemplifies a category of tools that I call *semimachines*—devices such as the blowpipe, the fire piston, the fire bow, the deadfall trap, and the snare—in which both the power and direction of motion of the device are imparted by the human operator. The potter's wheel, which also appears in the early years of Sumerian civilization (3rd millennium B.C.), is slightly closer to a modern concept of a mechanical device because its motion is more dependent upon its internal structure, but it still has only a single moving part and a single type of motion, and it can be classified equally well as a sophisticated semimachine or a borderline true machine. The security devices built into Egyptian pyramids are also borderline cases. The Egyptians closed the entrances of their tombs with heavy stone doors that slid down grooves carved into the walls on each side of the opening. Until the moment when the tomb was to be sealed, the doors were propped open with ceramic containers filled with sand. When the containers were shattered, the sand drained out of the grooves, and the doors moved into position, lowered by the force of gravity. But these ingenious security devices are functionally equivalent to deadfall traps, which are found all over the world among a wide variety of peoples with unequivocal folk technologies.

Although hand tools and muscle power are simple technology, they are not "primitive" technology. As archaeologists have confirmed, centrally managed teams of workers armed with implements are powerful forces in their own right; and two hundred years ago, the tools and techniques that built the pyramids were reinvented time and again to lift colossal statues from their resting places in the Near East and to transport them down to the sea for shipment home to Europe.[25] Heinrich Schliemann (1822–90), a millionaire businessman with a childhood ambition to find the site of Troy, conducted his excavations at the

Hellespont with a late Iron Age technology appropriate to the Roman Empire, which he faithfully recorded as if for a ledger: 150 workmen, 10 handcarts drawn by two men and pushed by a third, 88 wheelbarrows, and 6 horse-drawn carts. Although equipped with screw jacks, chains, and windlasses, all inventions of late antiquity, he was also well-provided with hand tools: 24 iron levers, 108 spades, and 103 pick-axes, "all of the best English manufacture."[26]

In the same way, a Cambridge don (Edward Clarke, 1769–1821) travelling in Greece at the end of the 18th century, was confronted with the problem of hauling away a two-ton marble statue from the ruins of the acropolis at Eleusis without the aid of wheels or pulleys. He had rope and nails that he bought in Athens, but the only tools locally available were an axe, a saw, and wooden poles. But as university graduates in those days considered it perfectly natural to order peasants about, especially ones who did not speak English, teams of laborers were soon assembled from a neighboring community, in spite of the fact that the people believed that disaster would befall their crops if the statue were moved. At the urging of the local Turkish official, the village priest was induced to turn the first spadeful of earth while wearing the vestments for High Mass; and the populace thus reassured, a "hundred peasants" were collected from the neighborhood of Eleusis and "near fifty boys."[27] Together, those in front pulled the colossus along with ropes, while others inserted rollers under it as it moved forward. Virtually the same method of transporting megalithic sculpture is depicted on an Egyptian mural of 1800 B.C., showing a statue about 25 feet high and estimated to be over 50 tons in weight being hauled along on a sledge by teams of workmen. In a study of quarrying techniques from ancient times to the present, one scholar concludes that the hand-held hammer and punch were not only the mason's tools of antiquity but that they showed "astonishingly little advance in the actual process of quarrying and handling stone until barely a century ago."[28]

That physical materials do not in themselves adequately characterize the technical complexity of a society is further corroborated by the history of machines themselves, which in Renaissance and Medieval times were made almost entirely out of wood and organic materials. For example, Agricola, the 16th-century chronicler, describes the water pumps which he saw in operation in mining areas of the Carpathians and the Tyrol. One type, the rag and chain pump, lifted water up a pipe by means of a spherical piston made out of horsehair sewn into a leather covering. The sphere was pulled up a pipe by means of a chain, pushing water ahead of it. Multiple spheres were attached to a continuous chain, which was rotated by means of a pulley, so that as one spherical piston popped out of the top of the pipe, another was taking its place below. At Chemnitz, three such pumps were arranged in tiers so that each machine could pump the water raised by the one below it. Agricola says that ninety-six horses were required to operate this system of pumps, which could raise water from a shaft 660 feet deep. In spite of the complexity of the mechanical design, these machines were made primarily of wood. Pipes were elm logs hollowed out by burning, the way Melanesians make a dugout canoe. Axles were saplings stripped of bark and sanded smooth. Circular cams were made of planks fastened side by side and rounded off to the correct shape like the bottoms of barrels. Gears were wooden spokes radiating from a hub, like the wheels of horse-drawn wagons, and the parts were fastened together with lashings, pegs, and mortise junctions. The machines of preindustrial times must have creaked and groaned like an old ship riding out a storm, but a vertical water wheel is a very forgiving piece of machinery. In fact, as some experimentalist might want to prove, an effective mechanical infrastructure could probably be built with a *Paleolithic* tool kit of stone hammers, chisels, and knives.

Advanced Folk Technology

In the ancient world, many technical processes that we regard as intrinsically large-scale and industrial, such as the manufacture of iron and steel, were done by village artisans using hand tools. In Africa south of the Sahara, for example, the smelting and forging of both copper and iron were commonplace among nonliterate farmers and cattle herders the length and breadth of the continent; and the steel produced by these peoples was often of better quality than the everyday implements of medieval Europe. A good example of folk-tech metallurgy is iron-making among the Yoruba people of Nigeria, which in precolonial times was done by a single village of metalsmiths who exported their products over a wide area. The society was visited early in this century by an anthropologist who described the technology in detail, and I have preserved his present tense in the following description.[29] The village is located next to deposits of hematite (Fe_2O_3), an iron oxide ore, which is mined by open-pit methods down to a depth of 6 to 8 feet. The ore occurs in nodules weighing a few pounds each, which are roasted all night over fires of green wood. Women and children then pound it to powder using wooden mortars and screen it with basket-like sieves. On the river bank, the women dig holes about 2 feet deep, fill them with water, and then, standing in the holes, wash the pulverized ore by rotating it with water in calabash trays, like prospectors panning for gold. The lightweight impurities are carried away by this centrifuge technique, leaving washed granules at the bottom of the dish. Other women sluice the remaining granules until the water comes away clear. Then the washed ore is carried to the smelter, where the men reduce it to pure metal.

There are 11 thatch-roofed smelting huts in the village, each 25 by 16 feet, with a furnace 7 feet in diameter sunk into the floor of each house. The furnace is dome-shaped and made out of clay, and it is punctuated with ventilation shafts, a chimney hole, and a door. The smiths, who are all men, fashion eighteen long, hollow clay tubes, about 1½ inches in diameter and two feet long, which are baked in a potter's oven. These are inserted in the air holes of the furnace, one end open to the outside. The fire is started with the aid of hot coals, and the ventilation pipes are sealed in position with clay. As the heat of the fire escapes through the broad hole in the roof of the furnace, the weight of the atmosphere pushes air through the narrow ventilation pipes, equalizing the pressure and creating a forced draft. For all of its primitive appearance, this oven is technically a blast furnace, because it forces a blast of air over the molten ore. When the fire is hot enough, charcoal, as well as clinker from previous fires, is thrown into the smelter. After two hours, the flux is drawn off through a hole in the bottom of the furnace, and this process is repeated two more times to clean out impurities.

Then about five pounds of ore are inserted, along with clinker and charcoal, and this process is repeated about every two hours, over a full day and night, for a total of ten charges, increasing the amount of ore each time. Before each charge, the slag containing the impurities is drawn off through a hole at the bottom of the furnace. When the iron is deemed ready, the clay seals around the ventilating tubes are broken away, and the red-hot "bloom" of pure iron, about 70 or 80 pounds in weight, is pulled out through the door by a loop of green vine and cut into more manageable pieces with axes. But even with seemingly simple techniques, the physical process is only a small part of the technology; and as shown in a recent study of the Mande people by the anthropologist Patrick R. McNaughton, traditional African metal-working is firmly embedded in a complex system of social transactions.[30]

The Power of Folk Technology

The technology of hand tools and village artisans, when organized into large political units, is a formidable source of power, and as recently as two hundred years ago even the armies of urban societies moved by foot and fought with hand-held weapons. Not surprisingly, time and again in the course of history, non-literate peoples introduced devastating military innovations to the arsenals of their times. Take, for example, the composite bow, which was probably first intensively employed, if not actually invented, by horse-borne raiders of the Asian steppe. Physically speaking, the bow is a borderline machine, a transducer that takes slow muscular force and converts it into speed, but the discovery of the basic principle is ancient.[31] Simple bows carved from a single piece of wood, with a waist in the center to fit the hand, are known to have been in use at least 6,000 years before Christ because they have survived intact in the peat bogs of northern Europe. However, the bow is probably much older than this, for flint points interpreted as arrowheads are known from some of the earliest Indian sites in North America, as well as from the Near East itself, among the hunting cultures that preceded the development of the agricultural communities in those regions.

The compound or composite bow, however, is a far more sophisticated device than the single-piece, or simple, bow, because it is constructed out of laminations of horn and sinew that give it much greater strength and resiliency. The horn and sinew are attached to opposite sides of the wood, with the sinew facing away from the archer. When the bow is bent, the sinew layer is stretched like a rubber band and tries to contract, while strips of horn, glued to the opposite face, are compressed and try to expand. The two forces together develop a strength that is beyond the elasticity of wood working alone. In fact, even to string a composite bow can be a feat of strength, and as recounted in *The Odyssey*, Penelope's suitors were quite unequal to the task. In the hands of a competent archer, mounted on a light chariot drawn by fast horses, with a teamster to handle the reins, the composite bow becomes a formidable fighting machine that can outmaneuver units of foot soldiers and overwhelm conventional cavalry with superior firepower.

The vehicle pulled by domestic animals was long known in the Near East, and both ass-drawn sledges and ox-drawn wagons have been found in the "Royal" Graves at Ur, dated around 2500 B.C. But the ultralight chariot with spoked wheels was probably pioneered somewhere to the north of the civilized Near East by a militarist people who combined it with horses and the armor-piercing bow to create a new assault technology that caught the ancient Egyptians napping. The kingdom of Egypt had natural defenses, the Mediterranean Sea to the north and the deserts of Libya and the Sinai protecting its flanks, but once these barriers were breached the Nile Valley was a highway that allowed the invader to advance unimpeded from one end of the country to the other. The Nile River runs down the middle of a flat plain, only 30 miles across at its widest, with steep bluffs of eroded sandstone on either side surmounted by arid wasteland stretching away to the horizon. Inside the corridor, the land is green and rich, and in ancient times, it was dotted with densely packed villages and punctuated by ceremonial centers, imposingly built of brick and stone. Ancient Egypt had an army composed primarily of infantry that moved to the battle on foot, followed by oxcarts with solid wooden wheels that were reliable but far from swift. But around 1700 B.C., warriors mounted on chariots and armed with compound bows, whom Egyptian chroniclers call the Hyksos, invaded from the Sinai and conquered the kingdom on the Nile.

The Hyksos ruled Egypt for a hundred years or more from huge, fortified camps, surrounded by high earthwork walls, in some cases 400 yards on a side, from which the charioteers would sally forth to collect tribute from the neighboring communities. They were eventually overthrown in a rebellion and driven all the way to Syria and across the Orontes and Euphrates rivers, and there is no doubt that the Egyptian victory was based on adoption of the new weapons. Chariots are not mentioned in the literature of the preceding Middle Kingdom, but tombs of the 18th Dynasty have yielded chariots of light and elegant construction, with the new spoked wheels, and the Egyptian words for chariot parts are borrowed from Syrian languages. In the Theban tombs of the 18th Dynasty, which followed the liberation, compound bows are buried with the pharaohs, and the walls show the princes getting lessons in archery. There is also evidence that the compound bows of the Hyksos invaders were armor-piercing weapons, for Pharaoh Amenhotep II boasts that he alone, of all men anywhere, could drive an arrow through three inches of Asiatic copper. But then Amenhotep II, who added to his harem in one year 270 women taken from foreign princes, was a man to be reckoned with on many fronts.

Amenhotep II's claim that he could drive an arrow through copper armor has been confirmed by anthropological research, which attests to the effectiveness of ancient weapons. Early in this century Saxton T. Pope tested many of the bows in the ethnological collections of the University of California and experimentally confirmed the armor-piercing properties of the composite bows from the time of Ramses III. Pope, a surgeon by trade, first became interested in archery after meeting Ishi, an American Indian who was taken to the University of California in 1911 after the rest of his tribe had been exterminated. Ishi is quite famous in anthropological circles, the subject of books and articles, and he taught archaeologists much of what they know about the manufacture of stone tools. His influence is also felt in the scientific study of bows, which he made for Pope's edification, and the two men even went hunting together in the wild. Pope later put his theories to practical test by hunting grizzly bears with a bow and arrow, and since he lived to write about it, his monograph on ancient weapons must be accorded credibility. Pope found, for example, that the English longbow, the weapon that decimated the French knights at Crécy, could drive a metal-tipped wooden arrow through a coat of mail or plate steel, demonstrating that the bow and arrow is a significant weapon; and he discovered that obsidian arrowheads are 25 percent more effective in penetrating unarmored flesh and hide than steel ones. Pope also built his own composite bow, sawing cow's horn into strips ½ by 12 inches, which he glued on a 4-foot-long hickory base that had been flexed by heating; and he reinforced the back with a hundred strands of catgut and a thin strip of rawhide. In this, he was following a published description of an ancient Egyptian bow of Hyksos inspiration, found in the Theban tombs and dating from the reign of Ramses II. Pope's replica had a pull of 85 pounds and could reliably shoot over 250 yards.

Five hundred years after the Hyksos, around 1200 B.C., at the beginning of the Iron Age, the Mediterranean world succumbed to yet another invasion by folk-tech armies from the north, a mass migration of what archaeologists call the Sea Peoples, after the terrified description by an Egyptian scribe: "These peoples were united, they laid their hands upon the countries as far as the circle of the world, the people which came from the islands in the midst of the sea." The folk societies were on the move again, and the whole political and economic framework of the Mediterranean world would be irrevocably altered.[32]

The Sea Peoples overran the Hittite fortresses in what is now Turkey and swept southwards along the Syrian and Palestinian coast. These invaders, the Philistines of biblical narrative, traveled in oxcarts with their women and children, while warriors in sleek, oar-powered boats, like the Vikings of a later time, raided up and down the coast of the Mediterranean Sea. The Egyptians first encountered them in large numbers in 1230 B.C. when they tried to invade from the west, allied with coastal Libyans, and Pharaoh Mer-ne-Ptah claimed to kill 6,000 of them and take another 9,000 captive. His successor, Ramses III, around 1190 to 1185 B.C., fought three more campaigns against the Libyans, who were being pushed into Egypt by raiders attacking along the coast. Shortly thereafter the Sea Peoples attacked Egypt directly, by a land assault through Palestine and by a seaborne invasion into the channels of the Nile Delta, depicted on an Egyptian relief from the period. The Egyptian military, counterattacking on two fronts, drove the invaders back before they could enter Egypt proper, but these wars and migrations mark the end of Egypt as a major power in the ancient world. The smelting and forging of iron, long promised as the metal of the future, finally became economical during this period of warfare and political turmoil, and it spread rapidly in the wake of the migrants and refugees, helping to undermine the supremacy of the imperious Bronze Age states.

As shown by the example of the Hyksos, with their powerful bows and horse-drawn chariots, and the Sea People with their fast oar-boats, tools and weapons made by village artisans were a significant political factor in the ancient world, and populous societies with homemade tools could often bring superior force to bear at weak points in the defensive cordon that surrounded agrarian nations. The foot soldier and the cavalryman, the archer and the charioteer, these composed the armies of antiquity, and folk technologies clearly had a fighting chance.

Not even the invention of the infantry firearm significantly altered the balance of power. Until the beginning of the 19th century, firearms were cumbersome weapons that delivered far less firepower per unit of time than the bow and arrow. They were developed as armor-piercing weapons, as a more effective alternative to the crossbow, but they were only more effective weapons when directed against troops wearing body armor of steel plate. In the frightening clamor of battle, the early musket must have been as vexatious as it was reassuring. In order to fire one of them, it was necessary to load priming powder into the pan; tamp powder, ball, and wadding down the barrel with a ramrod; lift the gun into firing position; and then touch off the primer with a length of smouldering punk. The military manuals of the 17th century list nineteen commands for loading a musket and fourteen more for firing it. The guns were heavy too, constructed of hardwood and steel, and they had to be rested on a staff which the soldier also carried around with him. The loading and firing took so long that soldiers had to fire in shifts, necessitating a high level of training and close-order drill, and they were vulnerable to traditional weapons while in between rounds. For this reason, the preindustrial musketeer wore a sword and body armor as well. The guns were also inaccurate, and rifled barrels did not come into use until the end of the 18th century. A well-practiced archer, in contrast, could fire an arrow about every ten seconds with a greater degree of accuracy.

Nor did the invention of mechanical weapons make the sword and buckler obsolete. Although the Roman army possessed sophisticated siege machinery, described below, that was beyond the capability of folk technologies to produce, these engines were developed for the attack and defense of fortified cities, not for guerrilla wars waged in the bush; and the compound bow of the Asian "barbarians" and the yew-wood longbows of the German "savages" were armor-piercing weapons, highly effective against the Roman infantry. Within their territorial borders, civilized societies could support large populations and maintain political hegemony over regions hundreds of miles wide, but even nonliterate societies with folk technologies could mobilize large numbers of men and bring them to bear with local superiority of force before reinforcements could be marched to the battle. For this reason, the empires of the ancient world surrounded themselves by walls.

The Chinese farmer of the 2nd century B.C. took refuge behind a stone wall 30 feet high and 1,500 miles long, while the Romans employed a similar strategy, a line of forts 5 to 20 miles apart, that ran from the Atlantic coast of Belgium to the western edge of the Black Sea, from the eastern shore of the Black Sea to the Red Sea, from the Nile River to the Atlantic coast of Morocco.[33] Across northern Europe, from the middle Rhine to the middle Danube, the forts were connected by a wooden palisade about 9 feet high, constructed of logs about a foot in diameter set vertically a foot apart, and the fence was positioned inside a steep-sided trench about 3 to 4 feet deep. Watchtowers and signal posts were constructed every Roman mile. Where the Rhine and Danube flowed wide and deep, the rivers themselves formed the frontier barrier, and forts were constructed on the fords, while watchtowers were positioned along the banks. Across northern Britain, where there are no natural barriers, the Emperor Hadrian (117–138 A.D.) constructed a stone barricade 73 miles long and 15 feet high, from the Solway Firth in the west to the mouth of the Tyne in the east. In front of it, facing the fierce Caledonians, was a parallel ditch 9 feet deep, while behind the wall was the Vallum, a double system of earthwork ramparts, each mound 6 feet high and 20 feet wide, separated by a ditch 10 feet deep. Fortified barracks, the milecastles, with housing for thirty to fifty men, were placed at one-mile intervals. In between each milecastle were three turrets, equally spaced, that were probably used as signal towers. At larger intervals were forts and fortified gates controlling access to bridges across the rampart system. Later on, the Romans built a second wall further north, between the River Clyde and the Firth of Forth. Only the stone fortifications of the Romans have survived to the present day, but Hadrian's Wall is merely a more substantial version of a system of watchtowers, ramparts, and legionary forts, interconnected by roads and signal posts, that was put into place along the entire 5,000-mile border of the empire, the northern and eastern lines strongly fortified, with a sparser presence in North Africa. What better measure could we have of the effectiveness of hand-tool technology than the extraordinary precautions that the Roman and Chinese empires felt compelled to take against it?

Machines Without Industrialism

In the industrial vision of the past, technical complexity, as measured by the number of parts in the artifacts or by the scale of the monumental architecture, is supposed to translate directly into political power and social change, transforming the society of which it is a part, but there is no better illustration of the fallacy of this idea than the history of machines themselves. Mechanical sources of power, intricate subassemblies, and the institutions of scientific engineering were all developed in the ancient world, a good two thousand years before the Industrial Revolution they supposedly brought about.

In 400 B.C. Dionysius the Elder, the ruler of the Greek colony of Syracuse, in Sicily, assembled his most talented craftsmen and mathematicians to prepare for a war against Carthage.[34] This was an intentional program of research and development, undertaken to create new weapons, and one of its most significant products was the catapult, a machine for hurling heavy missiles over long distances. The first catapult was a large bow and arrow, probably using the compound bow, made of laminations of wood, bone, and sinew, but one too powerful to be cocked by a man working unaided. Consequently, the inventors rested the arrow on a wooden slide that could move back and forth inside of a wooden stock, and they installed a metal claw on the slide that hooked under the bowstring like the tip of a finger. As the slide was pulled back, it engaged ratchets on the back of the stock that resisted its forward motion. The arrow was then fired by a trigger that pulled the claw, releasing its hold on the string. Before long, the devices were built so large that the stock had to be mounted on a stand and the slide pulled back with the aid of a ratchet winch.

The earliest catapults designed by the Greek engineers fired arrows with such force they penetrated both shield and breastplate, nailing the unfortunate victim to the ground; or even more spectacularly, impaled several soldiers with a single shot. The catapult was also used to hurl stone balls by substituting a slingshot pouch for the arrow, and the mechanical precision of the device ensured that successive missiles could be delivered to exactly the same spot, battering holes in city walls and knocking off the battlements one by one. Experimental models of some ancient catapults show that they were so accurate they could split the shaft of the previous arrow. Cities countered by rebuilding the walls of their fortifications to unprecedented thickness and established their own catapult batteries atop high towers so the defenders would have the advantage of greater range. This in turn led to a counter-cycle of technological innovation, an arms race to design ever more powerful catapults, and engineers soon reached the limits of the compound bow. So in place of the bowstring, they substituted a spring-driven device, called a torsion spring, which was made by wrapping stretched elastic fibers of horsehair around a wooden spool. With the torsion principle, fifty-pound balls and ten-foot arrows could be hurled hundreds of yards. Vitruvius, the Roman architect and engineer, gives specifications for machines capable of hurling stones weighing over 300 pounds. Although it is not known whether these huge machines were ever built, the standard field artillery pieces of the Roman army, which launched 57-pound balls, were substantial machines in their own right, massive wooden structures about three times the height of a man that were presumably assembled on the battlefield from a kit.

Greek engineers in the service of various rulers continued to develop catapult technology over a three-hundred-year period. Ctesibius of Alexandria developed two alternative power sources to the torsion spring that are harbingers of another age. In one version, tension on the bow arms compressed helical metal springs, made out of bronze, while in another version, tension on the bow arms drove a piston down an airtight cylinder, compressing the air and using its natural elasticity as a spring (and incidentally rediscovering the principle of the compression heating of air used in the Southeast Asian fire piston). But since neither of the new mechanisms was as powerful as the horsehair torsion spring, the engineers recorded these innovations as technical curiosities and concentrated their efforts on producing torsion spring motors with mathematical precision.

Around 270 B.C. in Alexandria and Rhodes, Greek cities ruled by the successors of Alexander the Great, the contemporary chronicler Philo reports that there were kings who fostered craftsmanship; and in these two cities, the construction of torsion spring catapults was made into a science in a strict sense of the term. The engineers devised a mathematical formula that expressed the dimensions of the spring needed to launch a missile of a given size. For the mechanized bow, the torsion spring bundle had to be ⅑ the arrow length for optimal results, whereas for a stone-throwing device, the diameter of the cord bundle in dactyls (the Greek word for "finger") is equal to 1.1 times the cube root of 100 times the weight of the ball in minas (a measure used for precious metals). In other words, the Greek engineers did not just build catapults but they devised a mathematical model that specified the power requirements in terms of the size of the missile, and they expressed the dimensions of all the parts as a multiple of the diameter of the spring.

Further evidence for scientific engineering in the ancient world can be found in Hero's *Treatise on Pneumatics*,[35] written in Alexandria around A. D. 62, which describes dozens of examples of complicated machines, many of them automata operated by inanimate sources of power. There are statues that pour water on an altar when a fire is kindled, a trumpeter that blows a bugle by means of compressed air, a coin-operated machine that dispenses drink, a vessel that remains full even though water is drawn from it, and so on. There are even steam turbines that raise the door of a temple automatically. Nor were ancient machines mere luxuries or curiosities. The Roman army was so heavily dependent upon machinery in the form of catapults and siege engines that a comparable number of artillery pieces would not be assembled on a battlefield again until the time of the Napoleonic wars.

The Roman army also pioneered the large-scale use of industrial prime movers—machines that power other machines—in the form of the vertical water wheel. Reconstructed models of ancient water wheels, which were powered by a stream of water directed against the blades by means of an aqueduct or millrace, show that they produced about five revolutions of the mill for every turn of the wheel itself; and this is a substantial gain over the power of horses or slaves. One Roman mill excavated near Naples is estimated to have moved the millstones at 46 revolutions per minute and to have ground 150 kg of grain per hour—compared to 7 kg per hour for two slaves with hand querns. Far from being insensitive to the benefits of machinery, the Romans understood power in its various forms as well as anyone before or since; and when the vertical water wheel became available, they put it to work in their characteristically big way. In the 3rd century A.D., at Barbegal, near Arles on the Mediterranean coast of France, they fed the waters of several streams into a common aqueduct that provided the power for eight pairs of undershot water wheels, arranged in a series on a 30-degree slope. Each wheel was 70 cm wide and 220 cm tall, built of wooden spokes on an iron axle. The water wheels were connected to horizontal millstones through wooden sprockets mounted in lead. This factory—why should it not be called a factory?—is estimated to have produced 28 tons of flour every ten hours, sufficient for a population of 80,000. There is also literary evidence for another facility of equivalent magnitude in Burgundy which provided flour for Roman troops stationed in northern Gaul. In ancient China too, machines and factories were widely used, and there is textual evidence for mechanically powered iron mills in the 4th century B.C. that employed a thousand people.[36]

The invention of machines, the use of mechanical power, the development of mathematical engineering, and the mechanization of subsistence tasks all preceded the Industrial Revolution by at least two thousand years. The *Domesday Book*, compiled by Norman conquerors of England in the 11th century to inventory their new possessions, lists almost 6,000 mills in England alone, and many of these were doubtlessly powered by water wheels.[37] Even as early as the 13th century, the European love affair with machines had evolved into what historians have called the "First Industrial Revolution." Water wheels were harnessed to a variety of crafts and industrial processes, such as sawing logs, pressing fruit for wine and cider, and grinding grain into flour. In the mountains of central Europe, mining for metals, first begun around 4000 B.C., resumed again in the 10th century after a brief hiatus caused by the decline of the Roman economy, and inanimate sources of power were widely employed to crush ore, control flooding in the shafts, and operate bellows for ventilation. In "preindustrial" Europe, the mechanically powered grain mill was as essential to society as the electrical generator is today, and historians estimate that there were 500,000 rural mills by the end of the 18th century—or one machine for every 29 people.[38]

The Display of "Fire Power"

Although human beings everywhere have technical skills and use physical objects that act upon the environment, the idea that these aspects of human life form a natural evolutionary unit whose progress and development can be chartered independently of the natural environment on the one hand and the cultural context on the other is precisely where the ideology of industrialism parts company with descriptive anthropology. Technological leadership is a mythical concept, not a technical one. Tools and techniques are judged to be technologically advanced when they exemplify the imagery of the charter mythology itself, either by rapid evolution from a simple prototype to powerful, complex descendants or by the development of huge, fire-breathing machines that emulate the golden, noble, head of the sun king (Nebuchadnezzar) himself. Computers, for example, demonstrate the technological leadership of the United States because they exemplify both the forward motion and the cephalic imagery of the myth. Not only did they evolve in the course of only forty years from room-sized aggregations of vacuum tubes to integrated circuits smaller than a thumbnail, capable of executing millions of instructions per second, but they emulate the faculties of logic and deductive reasoning which traditionally are localized in the brain and head.

As the historian George Basalla points out, the theory of technological progress is based on six assumptions:

> "First, that technological innovation invariably brings about a marked improvement in the artifact undergoing change; second, advancements in technology directly contribute to the betterment of our material, social, cultural, and spiritual lives, thereby accelerating the growth of civilization; third, the progress made in technology, and hence in civilization, can be unambiguously gauged by reference to speed, efficiency, power, or some other quantitative measure; fourth, the origins, direction, and influence of technological change are under complete human control; fifth, technology has conquered nature and forced it to serve human goals; and sixth, technology and civilization reached their highest forms in the Western industrialized nations. These assumptions are encapsulated in the well-known example of the advent of steam power."[39]

To the anthropologist, the steam engine, whatever its practical applications, is first and foremost a ritual artifact that expresses the progressivist theory of history so aptly summarized by Basalla—and it expresses it so well because it was designed to do exactly this.[40] Historical research reveals that there was a progressive increase in the efficiency of steam engine designs in the course of a 150-year period, as measured by the number of pounds of coal burned per unit of heat produced.[41] In 1712, the Newcomen engine consumed 32 pounds of coal per horsepower per hour; Watt's expanding steam engine of 1790 used 9 pounds; and the triple expansion marine engine of 1885 used only 1.5 pounds. This certainly looks like progress—but the appearance of linear progress is due to a careful juxtaposition of historical facts. For example, the first machine in this series, Newcomen's atmospheric pressure engine, is technically not a steam engine at all, because it is powered by the force of the atmosphere pushing a piston into a vacuum; and it is grouped with the steam engines by historians only because Watt was inspired by Newcomen's device. Yet without the inclusion of the Newcomen engine in the steam engine series, the graph of progressive efficiency would not be nearly as effective. Moreover, this progressive series implies that it was efficiency itself that motivated steam engine design; but efficiency only became a cultural value at the beginning of the 19th century, at the time of Trevithick and his steam-powered vehicles, and it was not a significant factor with either Newcomen or the younger Watt.

Rather, as one would expect in a ritual device, the imagery of power was initially far more important. Buchanan and Watkins, in *The Industrial Archaeology of the Stationary Steam Engine*, admit that they "cannot re-create in cold print the evocative smell of steam and oil or the impressively quiet throb of a moving steam engine." But they describe how Thomas Newcomen (1663–1729), an ironmonger from Dartmouth, England, took elements from existing machines—piston, cylinder, pump, lever, rocker arm, valve, and counterweight—and combined them with the force of air rushing in to fill a vacuum. Both vacuums and steam-powered devices were known to the ancient Greeks, but Newcomen's engine used the sudden injection of cold water into a steam-filled cavity to suddenly condense the steam and create a region of low pressure. According to the only contemporary account of the invention, the sudden cooling of steam was a serendipitous discovery. Newcomen had been experimenting with a vacuum engine, when a welded fracture gave way on the cooling jacket (which he used to condense the steam and create a vacuum), causing the piston to descend so rapidly and with such force that the machine was damaged.

Newcomen realized that the injection of cold water into the steam was far more effective than trying to cool down the cylinder, so he rebuilt the engine with the new principle in mind.

Almost certainly, it was the *imagery* of the engine that inspired Newcomen's dedication, for no hardheaded technological forecaster at the beginning of the 18th century would have put any money on the future of steam power. The containment of high-pressure steam was beyond the plumbing technology of the day, melting solder and bursting pipes, and physics had already demonstrated that vacuum-powered pumps could never be made practical. Only half a century earlier, Torricelli had proved experimentally that siphons, which he explained as the result of the weight of the atmosphere pressing down on a column of liquid, cannot raise water higher than 32 feet at sea level, and in mountainous regions even less—which meant that both atmospheric pressure engines and vacuum-powered water pumps were laboratory curiosities without practical application. If Newcomen had paid more attention to what had been scientifically established about steam power and vacuums, he might never have bothered with his engine at all.

But instead Newcomen forged ahead. He took the old idea of a piston and cylinder, added the method of creating a vacuum in the cylinder by condensing steam with a jet of cold water, and then used the force of the descending piston to move the handle of a reciprocating water pump. To top it off, he attached a counterweight to the piston, so that as steam refilled the chamber, the piston returned to the top of the cylinder to begin another cycle. That is, where Newcomen's predecessors had tried to suck up water directly into the low-pressure chamber, he used the small force of atmospheric pressure to tip a counter-weighted rocker beam attached to the handle of a reciprocating pump—transcending in one conceptual stroke the 32-foot limit imposed by the laws of physics on siphon-activated pumps.[42]

In spite of its originality and commercial success, Newcomen's invention was not generally perceived by his contemporaries as radically new technology. The basic mechanical concepts used in his engine go far back into antiquity, two thousand years or more, and the mythology of the industrial prime mover was still a hundred years in the future. Experimentation with steam and vacuums had consumed the collective efforts of scientists and engineers since the end of the 16th century; and in England, by the standards of the time, there was a massive investment by the state in basic research into pumps, metallurgy, and related technology, with an eye to developing more effective cannon and new, more powerful projectiles. The anthropologist Anthony F. C. Wallace summarizes these developments: "We can say that the steam engine was developed by persons for the most part employed by or associated with a single institution, His Majesty's Office of Ordinance, over the course of nearly a century, particularly that part of the ordinance establishment that was with its cadre of Dutch artisans located in and around Vauxhall in Lambeth. At the end, of course, the inventors had a little help from their friends in the Royal Society of London, and, at the very end, from an ironmongery business in Dartmouth."[43]

So it is not suprising that Newcomen's contemporaries did not see his invention as a great technological departure. The physical plant of industrialism, in the form of pumps and bellows, winches and cranes, trip hammers and crushers, was already in place at the mine and the forge centuries before the invention of the steam engine; and from a technical perspective, the development of steam power required only the substitution of a boiler for a water wheel on the handle of a reciprocating pump. To the people of the early 18th century, who did not share our progressive model of human evolution, Newcomen's atmospheric pressure engine was an incremental improvement in the water pump that would allow the mines to stay open all year round. Indeed, it was so little noticed by his contemporaries that no adequate biography of its inventor can be written because so few facts are known. Newcomen's is the sort of life that sets historians clucking over a line in a church registry that shows he delivered an order of nails.

Moreover, the prototypical invention of the "industrial revolution" is far closer to the definition of what humanistic critics of the industrial system call "appropriate technology" than it is to "high tech" as we know it today. The atmospheric pressure engine could be built from local materials by ordinary members of the building trades, using elm wood pipes and a glorified teakettle. Once constructed, the engines worked faithfully, day in and day out, for fifty years or more. They were not subject to the vicissitudes of weather, as were wind and water power, and they consumed a resource that at the pithead was dirt-cheap. Newcomen's engine even fits the social definition of appropriate technology for it was local rather than national, artisanal instead of bureacratic. And it eliminated much dreary and exhausting work.

But its *imagery* was industrial.[44] Newcomen's contemporaries referred to coal-burning machines as *fire engines*, not as steam engines, and the regular rise and fall of a huge oak rocker beam, delicately balanced atop a red brick tower three stories tall, the dark column of carboniferous smoke streaming from the chimney, the hiss of steam from the snifting valve, the white puffs of water vapor alternating with the thud of an eight-foot piston in a shiny brass cylinder—these must have made a strong impression on people who had only seen water wheels. No wonder the only surviving picture is a post card, of the sort that local merchants must have sold to tourists.

The imagery of "fire power," not efficiency, was the major factor in the development of the Newcomen engine, and efficiency was only recognized after the fact. Although Galileo had defined mechanical efficiency in the 17th century as the ratio of the work done to the energy consumed, the industrial application of the concept had to wait until fifty years after Newcomen, when the English engineer John Smeaton was sent by the Royal Society to study the windmills of Holland.[45] The English wanted to know whether the technology could be used to pump water out of tin mines, where Newcomen engines could not be practically employed because they consumed prodigious amounts of coal. Smeaton eventually concluded that it made more sense to make the Newcomen engine itself more efficient, which he did through a series of incremental improvements, such as fixing leaky valves and employing larger boilers. With these improvements, many atmospheric pressure engines continued to be used well into the next century, long after Watt perfected the principle of expanding steam.

Significantly, James Watt (1736–1819), the inventor of the steam pressure engine, was not initially concerned with questions of efficiency either.[46] Watt was a maker of scientific instruments by profession, which in those days included clocks, compasses, and music boxes; and while he was employed at the university in Glasgow, one of the professors of natural history asked him to examine a working model of an atmospheric pressure engine. The model had been sent once to London for repair but had come back in no better condition than when it was sent. Although Watt had never worked with steam engines, he had grown up in an intensely technical culture (his father was a shipwright and master carpenter, his grandfather a teacher of navigation and mathematics), and while working on the model he became enchanted with the mechanical potential of expanding steam. When Watt looked at an atmospheric pressure engine, he saw a steam pressure engine trying to get out.

In the Newcomen engine, the steam does not do the work but is used only to create a partial vacuum in the cylinder when steam is rapidly condensed by a jet of cold water—which in turn allows the weight of the atmosphere to push the piston down into the evacuated chamber. Watt, however, became fascinated with the expansive power of steam itself, and he began to entertain alternative designs for the Newcomen engine. As he recounted fifty years later, the idea for a "separate condenser" came to him suddenly one day while he was walking on Glasgow Green, and the hand-sized model that he built from scraps of tin and his wife's sewing thimble did not look like a Newcomen engine at all. The rocker beam was gone, replaced by the direct coupling of the load to the piston. The injection of cold water into the steam-filled chamber was gone too, replaced by an air pump attached to the condenser vessel, which initiates the cycle by drawing steam from the hot cylinder into the cold condenser. The cylinder itself, with its close-fitting piston, is maintained at high temperature by a steam-filled jacket that surrounds it. Most important of all, air from the atmosphere, instead of rushing in to fill the low-pressure cylinder—thereby pushing the piston down—is now kept out by air-tight seals so that steam itself can do the work.

Watt's engine required a much higher standard of construction than the Newcomen engine. With the latter, all the parts, with the exception of the piston and cylinder which had to be made at a foundry, could be constructed by millwrights and carpenters using locally available materials. But Watt complained that even experienced foundry workers lacked the skills to produce parts to the tolerances he needed to contain pressurized steam, even though the steam he needed was pressurized at barely above an atmosphere. In fact, the commercial production of steam engines required mechanical machining of pistons and cylinders to ensure a snug fit while still allowing unimpeded movement; and Watt became one of the Wilkinson Brothers' first customers, buying cylinders milled on the boring machine they had developed to produce cannon barrels for the Royal Navy. Years later the Wilkinsons returned the favor by buying a rotary steam engine from Boulton & Watt to power the machines of their foundry.

Watt did not develop the principle of expanding steam "in order to" make more efficient engines, but, unlike an earlier generation of craftsmen, he did appreciate the commercial implications of engines that burned less coal. Moreover, he created a direct and observable link between the work done by machines and the amount of money saved and invested. Most collieries, with access to unlimited supplies of fuel, were not interested in more efficient coal-burning engines, and most continued to use their Newcomen engines, later retrofitting them with separate condensers.

So the firm of Boulton & Watt concentrated on tin mines, which had the same problems with flooding but no cheap supply of readily accessible fuel. Although the first commercial steam pressure engine was in fact erected at a coal mine, in 1778, tin mines formed the majority of early sites. Initially, the firm of Boulton & Watt wanted to charge its customers an annual fee equal to one third of the savings that accrued from using a Watt engine in place of a Newcomen engine, but this proved far too mathematical to the mine owners of the day. So Watt invented a meter that counted the number of strokes made by the piston, and he put it in a locked box attached to the engine. At the end of each fiscal period, a clerk from the company checked the meter and computed the amount of water raised. By this device, Watt demonstrated a systematic relationship between the efficiency of an engine and the cost of running it. With the rotary steam engine, introduced in 1783, the link was made even more observable. Since the new machine was not used to pump water but to turn mills, Watt computed its productivity by the number of horses it replaced. The available statistics showed horses to be very variable workers indeed, so Watt, with the British genius for weights and measures, arbitrarily set the horsepower unit at 33,000 pounds per foot per minute. Now, both living things and inanimate devices could be reduced to a common scale.

In his concern for objective methods of pricing, his development of mechanical devices to implement the billing process, and his use of engineering concepts to compute the efficiency of work, Watt crossed the boundary from traditional engineering to industrialism. As Ellul has emphasized, industrial technology is not characterized by particular kinds of machines—because the machines are always changing—but by the application of the concept of mechanism itself to the organization of work and to questions of social and economic policy. Machines and inanimate sources of power are more than two thousand years old, but only 18th-century Britain applied the engineer's concept of mechanical efficiency, epitomized by the steam engine, to the large-scale organization of commerce and industry. Efficiency had become a cultural value.[47]

This development was facilitated by the circumstances under which steam engines were used. At first, steam pressure engines were erected at the work site, as if they were a building, but the application of steam power to vehicles made the concept of mechanical efficiency of great practical concern. Unless the engines could be made more efficient, they would burn more coal than they could carry. Watt's successors, engineers like Trevithick, regarded reduced fuel consumption as a technical challenge, but Watt opposed this development, for he regarded high-pressure steam as far too dangerous to be used in everyday machines, which, in fact, exploded with some regularity through the remainder of the century. When efficiency becomes the standard by which money is invested and engineers evaluated, it is not suprising that the steam engine evolved from a coal consumption device to a coal conservation device in little more than a hundred years, its evolution made possible by the development of techniques to handle steam at ever higher pressures and to reuse the heat from previous cycles. But it is the practical requirements of steam-powered vehicles and the imagery of mechanical power itself—not any historical law—that underlies this asymptotic curve. In fact, as this brief historical synopsis shows, the progressive efficiency of steam engine designs is a kind of back projection of 19th-century values onto the process of history itself.

Advanced Heads and Primitive Behinds

As historians in recent years have shown, all the institutions of capitalist economies *preceded* the technical processes that supposedly brought them about.[48] By the middle of the 17th century, there were scientific institutions funded by the state and dedicated to pure research, military R & D facilities that applied scientific innovations to the development of weapons, financial institutions that invested in prospective sources of wealth, a commercial sector that owned productive facilities and managed them for profit, partnerships and joint stock companies for undertaking new commercial ventures, markets both domestic and international for selling and distributing new products, and taxation of commercial revenues to fund the cost of government—all of this a hundred and fifty years before the major technical indicators of the industrial revolution.

It is clear that the social relations of industrialism precede industrial technology as conventionally defined, but even among critics of progressive evolution, there is still incredible resistance to the idea that technology is not an objectively definable entity whose existence transcends the cultural contexts where artifacts are produced and used. For example, George Basalla, in *The Evolution of Technology*, published by Cambridge University Press at the late date of 1988, argues convincingly that progress, efficiency, and necessity are inadequate explanations of technological development; and he criticizes Eurocentricity, unilinear development, and uncritical extrapolations of biology to culture. But his machines are nonetheless connected to each other as ancestor and descendant, divorced from the social and ecological context, and arranged like specimens on a museum shelf. When all is said and done, this author shows no awareness whatsoever that the very act of demarcating something called "technology" that exists "outside" of nature with its own "evolutionary development" and which can be represented as "species" of machines arranged in "chronological order" is just the progressive history of the steam engine all over again.

In the same way, archaeology, which has contributed so much to our understanding of the role of multiple pathways in technological change, has also continued to perpetuate the idea that technology is tools and techniques arranged in chronological order. Industrial archaeologists, for example, define "industrialism" as the use of steam power in basic industries, while even the authors of textbooks in anthropology tell us that "ceramics" is an "ancient technology," which they illustrate with Navaho pots. Why don't they tell us that ceramics is a contemporary technology and illustrate it with superconductors and the tiles on the space shuttle? Although there is no doubt that over the past 10,000 years unprecedented innovations have appeared on the planet earth—geared machines and gasoline engines, calculators and telescopes, processed foods like bread and wine, textiles and ceramics—the common interpretation of these events is still Nebuchadnezzar's Dream with its stages of primitive and advanced. Yet this theory cannot explain the million-year plateaus that characterize the stone tools of the Paleolithic, the cycle of decline at places like Rocky Cape, or the faith in progress that subsidizes industrialism.

"Technological progress" is the incorporation of machines into Nebuchadnezzar's Dream, so that some countries are "a head" while others are "behind(s)," and this perspective looms so large in the modern field of view because we present the past as a series of snapshots of material remains arranged in chronological order. But if the camera is pulled back to reveal the social and ecological context of machines, then the clear-cut image of progressive development blurs into a patchwork of diverse technical traditions and culturally specific systems of imagery. Indeed, none of the commonly accepted propositions of technological development are confirmed by contemporary scholarship: technology does not develop from the simple to the complex; hand tools are not ancestral to machines; civilization is not a consequence of agriculture; writing is not a prerequisite to high technology; folk societies do not have primitive techniques; factories did not begin in the 18th century; specialization is not a measure of technical complexity; industrialism is not the mechanization of the means of production; and human life did not begin with small, egalitarian bands of hunters and gatherers. This universally accepted picture of the past is an inversion of a biblical myth.

BLOODY STUMPS, EXPLODING STARS

The Iconic Logic of Purified Bodies

In all societies, technology is the ritual replication of origin myths such as Nebuchadnezzar's Dream, but industrialism regards myth and ritual as primitive modes of thought that need to be replaced by bureaucracy, science, and machines. How does industrialism reconcile the social necessity of myth and ritual with its own technocratic agenda? It does this by operating on two distinct levels of symbolic structure simultaneously: an overt level of mechanism and rationality, which is openly acknowledged and taught in the schools as a true description of the world, and a covert level that is officially dismissed as mere "art and entertainment" but which in fact creates a mythic infrastructure supportive of the technocracy.[1] Thus, the symbolic body of industrialism has three distinct categories of symbolic content: the "primitive" aspects of nature and society that the system is trying to replace (equivalent to the feet of Nebuchadnezzar), the technological "advances" with which it hopes to replace them (equivalent to the head of Nebuchadnezzar), and the covert system of myth and ritual that defines Nebuchadnezzar's Dream itself. However, in the folk theory of industrialism, Nebuchadnezzar's Dream is equated with the primitive parts that the technocracy is trying to eliminate, so it cannot be openly acknowledged but manifests itself by what I call *covert symbolism*.

The overt content of industrialism is conveyed by a broad range of so-called educational institutions, including schools, museums, and documentary television. It is also taught and reinforced by the business and financial communities, both through their own media and by dissemination in the general press of statements by stockbrokers, investment strategists, and economists of all persuasions. The "primitive" parts of industrial society are obvious too, as commonplace as churches, religious schools, family gatherings, and company picnics. However, the covert content of industrialism is a good deal harder to recognize because nowhere is there anything that says: "Covert mythology: read me!"; "Hidden agenda: do this!" In fact, the culture actively denies that there is any such thing, proclaiming in large letters: "No myths here!"

Nonetheless, industrial society has several categories of covert myth and ritual, each conveyed by a characteristic genre or medium. First, there is the ritual of the technocratic extravaganza, typified by the space shuttle launch and the lunar landing, through which ordinary citizens are incorporated into the activities of high-tech organizations. Like rituals, these carefully orchestrated events take place in real time with real characters, incorporating people from all walks of life into the enterprise as spectators. Yet these events are also art, for even when broadcast "live," they are narrated by newscasters, edited by control-room technicians, and interspersed with previously recorded footage. The space launch, by merging the imagery of the technocracy with that of the adventurer/hero, and by referencing the great American aviators and explorers of the past, such as Charles Lindbergh and Admiral Byrd, makes it possible for people without any technical knowledge to understand the purpose of the activity and to identify with its goals. Moreover, the ritual battle between "man and nature" has inherent drama, for as the explosion of the space shuttle *Challenger* showed, there is always the chance that "nature" might win.

In addition to these national rituals of technocratic dramatization, there is also the covert mythology proper, which is primarily conveyed to citizens of industrial society by the genre of science fiction (sci-fi to its fans); and it can take the form of books, toys, television shows, video games, and motion pictures. With a few notable exceptions, the books are rarely read by anyone outside of the sci-fi community, but the movies and TV shows, which are directed to mass audiences, often attain the status of cultural icons: the *Star Trek* television series and the movies it inspired; Stanley Kubrick's *2001: A Space Odyssey;* and, of course, George Lucas's *Star Wars*, which was for a time the largest-grossing movie in Hollywood history. Although the genre is called *science* fiction, the content usually has very little to do with the scientific method, scientific theory, or scientists themselves.[2] In many stories, the heroes are not even scientists, and in others the social organization, far from reflecting the intricate division of labor of technocratic society, often appears to be lifted wholesale from medieval Europe, with knights and warriors, peasants and aristocrats, priests and magicians. The modes of transportation often mix radically different levels of technology, with some characters riding on domesticated animals while others pass from one galaxy to another on warp-speed spaceships in a matter of minutes. If managers and bureaucrats appear at all, they are often portrayed as the tools of evil empires or as officious functionaries putting obstacles in the path of the hero. If commerce has any role in the plot, it tends to be an underground activity of pirates, bounty hunters, and unscrupulous cartels. The plots are more likely to involve physical challenges than intellectual ones, and many of the heroes scarcely differ from cowboys in terms of their physical stamina and combative proclivities. Yet, in the sci-fi genre, weapons are curiously primitive given the technological standards that prevail elsewhere in the plot, almost always being hand-held and optically guided, so that shots by bad guys rarely find their marks. Although scientific facts and theories are often incorporated into the story to make certain aspects of the plot more plausible, such as spaceships propelled by gravity waves, they are just as likely to be ignored, as when the hero travels faster than light or when objects are moved by telekinesis.

In *Star Wars*, for example, the villain is a vast bureaucratized empire staffed by Teutonic-looking generals and faceless storm troopers (the term storm trooper is in fact used in the spinoff books), represented by the emperor's most powerful warrior, the evil Darth Vader.[3] The heroes are a ragtag band of insurrectionists consisting primarily of a blond, young male of noble lineage (Luke Skywalker), a princess (Leia), a personable smuggler with a noble heart (Captain Han Solo), his alien sidekick (the Wookiee named Chewbacca), and a pair of androids, R2-D2 and C-3PO. The Empire commands vast technological resources, including a man-made planet called the Death Star, but the storm troopers are armed with weapons that look very much like 20th-century automatic rifles, and both the arch hero and the arch villain prefer hand-to-hand combat with sabers—laser-like *lightsabers* to be sure—but weapons that nonetheless require the physical skill of fencing. The nonrationalist side of science fiction is also well represented in this story by an old man in a monk's cowl, Ben (Obi-Wan) Kenobi (the parentheses are part of the name), who teaches Luke Skywalker about the Force. The Force is the power of life itself, and it is said to be more powerful than the mere technology of the Empire. Before the arch villain Darth Vader kills Ben Kenobi in a duel with lightsabers, the old man prophesies that he will become even more powerful in death; and, sure enough, after Obi-Wan is killed, his body suddenly disappears, leaving behind his monk's habit in a rumpled pile on the floor. In both its technological imagery and surreal fantasy, *Star Wars* is a good representative of the sci-fi genre, so one is naturally led to the question: Where is the science in science fiction?

Science fiction is *science* fiction because it encodes the symbolic content that is a necessary part of the social organization of the technocracy but which cannot be overtly acknowledged because it conflicts with the rationalist and mechanistic world view: the contradiction between scientists' perception of themselves as rugged individualists and the demands of the bureaucratic organizations which they invariably serve; the persistence of manual skills, which, if technocratic history were right, should long ago have been replaced by machines; the arbitrary and capricious use of political and economic power in what are supposedly rule-governed, democratically organized institutions; the political competition between individuals in what is supposedly a meritocracy; the fact that technical superiority has more to do with old-fashioned martial virtues than with the dispassionate discovery of truths about nature; and, finally, that there is some poorly defined spiritual power, the Force if you will, that is stronger than machines.

The fact that the industrial system produces two distinct symbolic products, the overt and covert symbolism, science and science fiction, indicates that there are two classes of technocratic practitioners, what I call the *mechanists* and the *players*. The mechanists comprise the vast majority of people who subscribe to the technocratic world view, and they believe that the overt mythology is an accurate description of reality: that people are machines, that labor is a quantity, that the past is primitive, the future advanced, and that they and their colleagues are engaged in rational activity in a world that can be made even more so. It is the mechanists who develop the algorithms, design the machines, perform the experiments, perfect the standardized tests, compile the statistics, edit the journals, program the computers, write the plans, compose the press releases, and do the million other functionally defined jobs required by the bourgeois conquest of knowledge. If the mechanists read science fiction, they dutifully think of it as a form of entertainment. If they watch a space shuttle launch, it resonates to beliefs about American technological leadership, the conquest of the last frontier, and the vindication of individual initiative. In short, to the mechanist, there is no covert agenda in industrial civilization; and even if there were, it has long been purged of its incense, icons, and relics of saints—like a pure and triumphant Protestantism.

However, there is another type of technocrat, whom I call the player, who recognizes that mechanism by itself just dead-ends in the middle class, and that the technocracy has both overt and covert symbolic content. The player understands, perhaps only on an intuitive level, that *Star Wars* is not just a movie but as essential a part of industrial technology as the space program itself. As Delany has pointed out, the distinguishing characteristic of science fiction as a genre is that it creates imaginary worlds that are to be taken as possible futures, in effect, a world without metaphor, in which even seemingly metaphorical statements, such as "His eyes popped out when she walked into the room," are always to be given physical interpretations. This literal-mindedness is a characteristic of the technocratic mentality, and it is reinforced by long exposure to its most characteristic machine, the computer, so when the player, who is as literal-minded as any other technocrat, and as lacking in social imagination, views *Star Wars*, he does not see it as most people do, as an adventure story, an allegory, or as a stunning display of special effects, for all these interpretations presume metaphor or indifference. Rather, the player's only possible interpretations are either skepticism or belief, so he must either dismiss the film entirely or see it as a literal description of someplace else or some other time, as actual events that have either happened in the past or will happen in the future. Even though science fiction is clearly labeled "fiction," not fact, and certainly not myth, the player nonetheless comes away with the conviction that the real technocratic agenda is not the machine but the Force: an essentially irrational and mythical domain of princesses, gurus, and Jedi knights.

The player's confidence is not misplaced because a movie like *Star Wars* is itself a high-technology product as sophisticated in its way as any computer system or spacecraft. In Alan Arnold's book, *Once Upon a Galaxy: A Journal of the Making of "The Empire Strikes Back,"* two of the principals involved in the making of the movie, the production designer and the co-supervisor of special effects, tell us that they keep abreast of the work at the United States government agency that manages the space program (NASA), and that this interest is reciprocated, for many NASA people are keenly interested in science fiction movies. Moreover, the production of the *Star Wars* sequel required state-of-the-art technology, including computerized image-processing that would not be out of place in the Jet Propulsion Laboratory, as well as the services of two special effects studios, one in London to work with the film crew, another in California, with a staff of forty-three technicians, to build the models. The sets for the movie also required an industrial level of technical organization. Han Solo's spaceship, the *Millennium Falcon,* was constructed by a firm of marine engineers at the Pembroke

Docks in Wales, where flying boats were built in the 1930's. This one prop, constructed of steel, was 65 feet in diameter and 16 feet high, with a weight of 23 tons. It was transported to the studio in London in sixteen separate sections, carried by a convoy of trucks. The other props included specially designed snowmobiles, built by a car manufacturer, and remote-controlled, working models of robots produced by a toy company. In addition to the sets and props, shooting the location scenes required a level of logistics appropriate to an airborne assault. For the rebel stronghold on the ice planet Hoth, Lucasfilms took over a glacier at Finse in Norway, where it erected two base camps, equipped with arctic survival tents and a full-time medical staff. Ten container loads of equipment went by sea from England to Oslo, then by train to Finse. Three others were airlifted in. To move the actors and crew between the two camps and the hotel, there were ten tracked snow vehicles, a snowplow, and a helicopter. Movies like *Star Wars* do not only portray technocratic values: they are themselves technocracies.

Because there is no hard-and-fast dividing line between industrial technology and the media representation of it, we can take the *Star Wars* trilogy as a projection of the repressed aspects of the technocratic culture that produced it. But as movies are themselves industrial products, sensitive to all the same taboos that operate elsewhere in the system, their symbolic content, especially that of large-budget extravaganzas, is never an unambiguous projection of the covert symbolism. Movies are subject to the rigorous censorship of the box office, and their symbolism must hint at the covert content without actually exposing it. For this reason, the covert content of industrialism and the surface projections of it are quite different, in no way identical, but they are nonetheless systematically related to each other through processes of symbolic transformation. In this sense, science-fiction movies are the dreamwork of the technocracy, the configuration of words and images, plot and characterization, that recreate the repressed world of ritual and myth in a form acceptable to the culturally dominant consciousness.

The anthropologist's task, like the clinician's, is to infer the underlying system of symbolism from its disguised manifestations on the screen. To do this, we first assume that the symbolic content is logical, systematic, and productive; and then we build a cognitive framework in which the events on the screen can be seen as coherently motivated and rational. The linguistic concepts of deep structure and surface structure are useful here. The surface structure is symbolic content that is directly observable, whereas deep structure is the inferred system of cultural rules that produces it. For example, in English, the list *horses, dogs, cats*... is surface structure, whereas the rules for pluralizing these nouns are deep structure.[4] Thus, in analyzing imagery and myth, the anthropologist develops what I call an *iconic deep structure*, or IDS: a system of imagery and rules of transformation that explains the observable features of the symbolism (Figure 3.2). Like any hypothesis, the iconic deep structure must, first, be consistent with the empirically observable content; and second, if the generalizations are correct, it should be possible to extend them beyond the original evidence to explain other works in the same tradition. Thus, *an IDS is a hypothetical system of thought that explains the collective dreamwork of a culture*—and I use this idea to explore the covert symbolism of industrial society.

As Sigmund Freud showed in *The Interpretation of Dreams*, information that is unacceptable to consciousness, because of cultural taboos or whatever reason, manifests itself in a distorted or disguised form, such as jokes, dreams, or "Freudian" slips, a process he called *repression*.[5] Freud's original insight was based on his experience as a medical doctor. As medical practitioners are well aware, people who lose a limb by amputation frequently have the experience that the limb is still attached to the body, a well-documented phenomenon known as the phantom limb experience. In this syndrome, the body part is objectively missing, but the patient's body, on some fundamental experiential level, refuses to accommodate to the altered morphology of its physical form. Freud's concept of repression is the logical opposite of the phantom limb effect, for in the repressed patient the body part is objectively there, apparent for all to see, but its existence is experientially denied, so that the person acts as if the body part were not attached to himself. Implicit in Freud's original formulation of repression is the idea that there is a systematic relationship between traumatic acts directed towards the body on one hand and the fantasy or denial of bodily experience on the other. Repressed patients no longer experience physically attached parts of their body, just as amputees fantasize body parts that are no longer there. Freud saw the source of these phenomena as trauma, a medical term meaning a wound or injury to the body; and he generalized it to include purely symbolic acts—mental traumas that leave the physical body intact but seriously damage the internal perception of it.

However, where Freud saw the process of repression as taking place primarily inside the head of the individual patient, I see it as a fundamentally social process that is defined in terms of the symbolic body of a culture.[6] In the symbolic anthropological approach, *repression* occurs

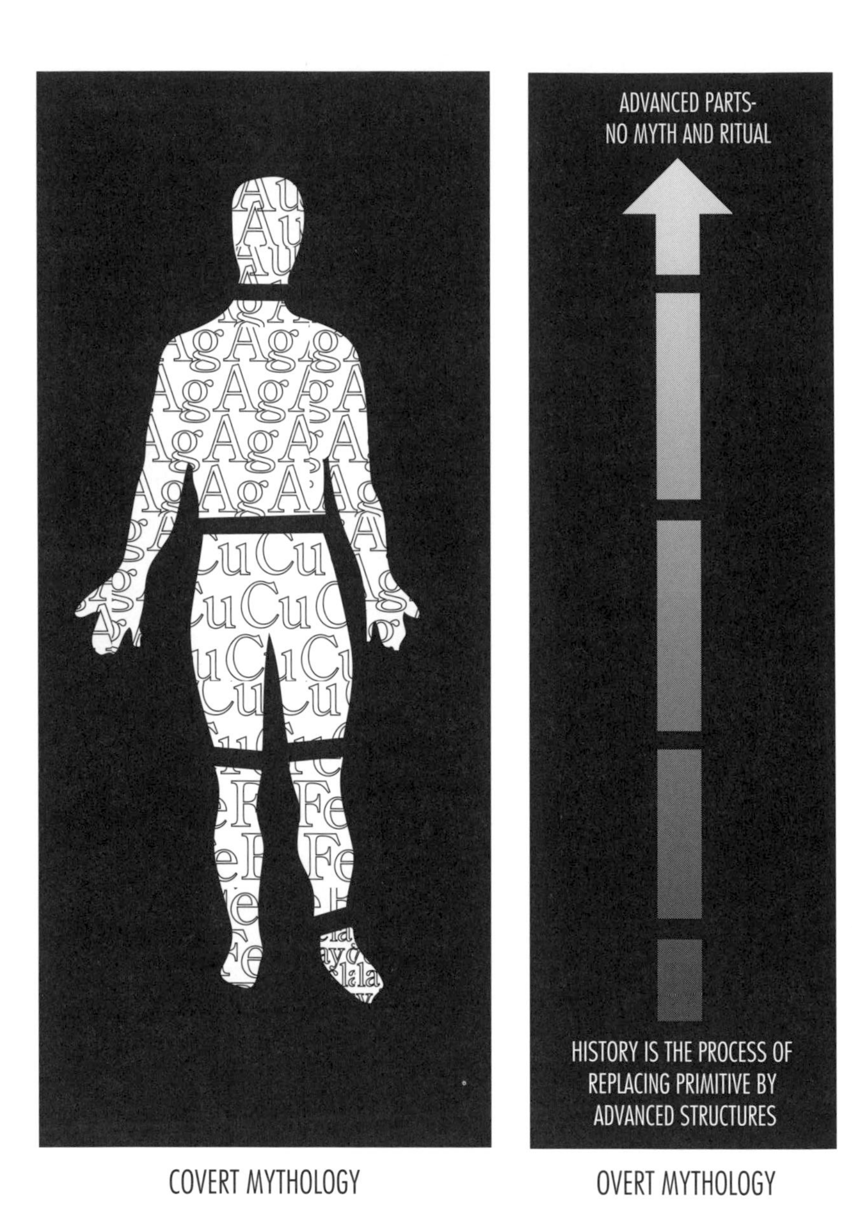

Figure 3.1 Folk theories of technological change are based on the covert imagery of the symbolic body.

when a culture collectively ignores or denies the relevance of parts of the human body and fails to represent them in its definition of the "human species." That is, in repressive situations, the culture establishes an idealized body image that differs in significant ways from the biological body, and it prescribes techniques, both physical and symbolic, for bringing the observable world into conformity with this ideal. Typically, the biological body itself is altered, by means of clothes, makeup, tattoos, deformations, and so on, to ensure that real human beings match the idealized image. In addition, in the realm of art and imagery, the taboo body parts are removed from the collective representation of the human species and relegated to an underworld of unconscious forms. Thus, repression is equivalent to an act of mutilation on the symbolic body that produces an amputated symbolic body part (called the *amputatum*) and a purified residue of the process (the *puritatum*). The purified residue becomes the overt part of the cultural representation—the official body image—whereas the amputata are relegated to the covert symbolic content.

The *Star Wars* trilogy is ideal for demonstrating this interaction of overt and covert symbolism. First, the movies were great commercial successes, indicating that they resonate to the values and beliefs of contemporary Americans. Second, they are representative of the genre, and Baird Searles says in his *Films of Science Fiction and Fantasy* that for many people *Star Wars* is the best science-fiction film ever made.[7] Also, they consciously present themselves as "high-tech" movies, both in the technology portrayed and in their method of production. Moreover, they spawned a whole host of spinoff products, including adventure novels and children's books that elaborate on the symbolism, making it more accessible and transparent. In addition, the movies are available on videotape, so it is easy for anyone interested in analyzing them to view them at slow speed and to stop at particular frames. Finally, the name "Star Wars" is symbolic enough to have entered popular culture in another way: It was adopted by critics of the arms race as a derogatory label for the proposed nuclear defense shield of laser-toting satellites first introduced by President Reagan and known to its advocates as the Strategic Defense Initiative or SDI.[8]

The Mythology of *Star Wars*

Oddly enough, the first film in the trilogy, *Star Wars*, released by Twentieth Century-Fox in 1977, begins with the subtitle "Episode IV:

A New Hope." Why does the series, especially a series with three movies, begin with Episode IV? Although George Lucas is on record as saying that two trilogies were originally written, not one, this does not explain why he started with the second half of the story or why he did not simply renumber the episodes. Such discrepancies in the overt symbolism are usually indicators of repressed content on a deep structure level, and the task of the symbolic anthropologist is to postulate an IDS that explains the anomalies on the surface. One interpretation that is consistent with the symbolic content of the movies is that the Roman numerals are to be taken as letters. In this case, the number *IV* is really a pair of letters, "I" and "V," which together form a commonly used abbreviation for "intravenous," a phrase known to anyone who has been in an American hospital. Venous blood is deoxygenated blood, dark red in color, that is returning to the heart, and *intravenous* means "inserted into the veins," as with a syringe or an IV catheter. It is consistent with this interpretation that at the very beginning of *Star Wars*, when Princess Leia is captured by the troops of the Evil Empire, she is injected with some sort of drug by a hovering, hypodermic robot shaped like a black sphere.

In addition to the image of foreign substances injected into the blood, the name of the arch villain, the black-robed and helmeted Darth Vader, is in my view a symbolic transformation of the phrase "dark invader," conceptualized as an intruder into the collective body. Significantly, in the spinoff books, he is denoted Lord of the Sith, where Sith is obviously an anagram of shit, so the reference to bodily functions is not far off. Freud showed in *The Interpretation of Dreams* that symbolic content is often disguised by anagrams that rearrange the letters of words, by puns that substitute one meaning for another, by the deletion of speech sounds or letters, and by the combination of separate words into new condensed images. As shown in Figure 3.2, all of these processes are at work in the transformation of the phrase Dark Invader into the name Darth Vader. In English, the phrase "Bite your tongue!" is used when someone says something that they ought not to have said, and this gesture is also a speech sound in the language, ordinarily written "th," and articulated by placing the tip of the tongue between the teeth, as in the last phonemes of *tooth* and *teeth*, and the first phonemes of *thick* and *thin*. Linguists call this speech sound theta, written θ, and if you apply the cultural injunction "Bite your tongue!" to the word *dark* as you pronounce it, the final consonant is transformed from k to θ—from *dark* to *darth*. In addition, the dreamwork removes incongruous elements from *dark invader*. For example, if the phrase *dark invader* is run together as a single word (*darkinvader*), then the word *kin* is seen to be embedded in the middle of it. But as consciousness recognizes that this element is inappropriate to the larger meaning of the phrase, indeed conflicts with it, the dreamwork says "No kin of mine!"—and out goes *kin*, transforming *darkinvader* to *darvader*. Indeed, Darth Vader's audio anxiety is even audible in the movie as the heavy-breathing track that follows him around like an obscene caller. These two transformation processes together produce the name Darth Vader from the prototype *dark invader:*

darkinvader + No kin of mine! → *darvader*
and
darvader + Bite your tongue! → *darthvader*.

In confirmation of this interpretation, Darth Vader's name before he became the evil servant of the Emperor was Anakin. This name not only incorporates the *kin* element extracted with the first transformation, but *ana* itself is a prefix with several meanings, one of which is "similar to," as in the word *analog*. So Anakin translates to Darth Vader in his kin-like aspect: that is, Darth Vader as Luke Skywalker's father, a genealogical fact that is revealed in *Return of the Jedi*.

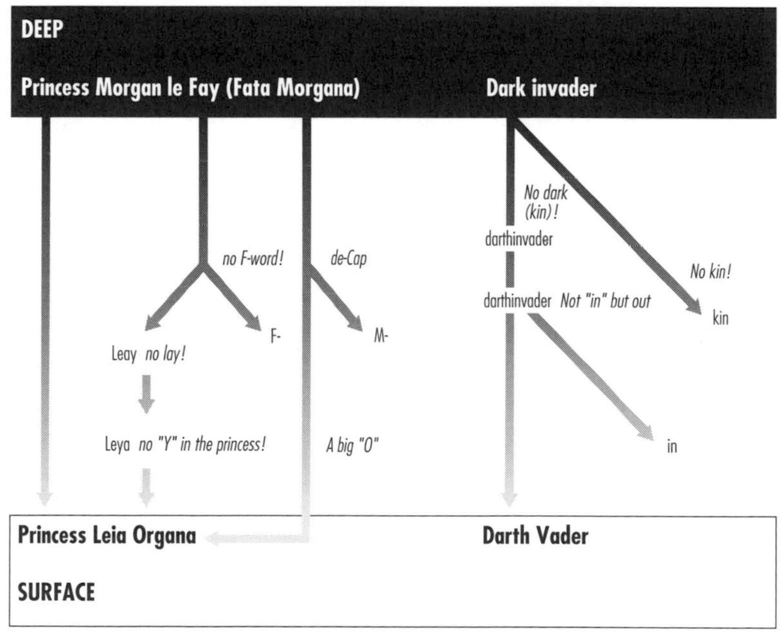

Figure 3.2 The names of characters in *Star Wars* are transformations of covert imagery.

But if *kin* combines with *ana* to become *anakin,* then where does *ana* itself come from? The title Lord of the Sith suggests *anal* as a possibility, but interpretations that explain multiple elements in the story are preferable to those that explain but a few. For this reason, I suggest that the first half of *Anakin* is primarily derived from *Fata Morgana,* the elusive mirage seen in the Strait of Messina and attributed to King Arthur's wicked half sister Morgan le Fay. This too is not surprising, for there is already a princess in the story, and princesses imply kings. Also, the term *morgan* has strong genealogical associations that go well with *kin,* encoding both a blood relation of the mythical King Arthur, and a form of marriage, namely *morganatic,* which is a union between a noble person and one of inferior rank in which neither the lower-ranked person nor the children born of the marriage are eligible to inherit the noble's title or property. The hypothesis that Morgan le Fay is the source of both Anakin and Princess Leia is confirmed by the credits of the movie which give Princess Leia's full name as Leia Organa. As shown in Figure 3.2, the prototype phrase "Princess Morgan le Fay (Fata Morgana)" is transformed by a series of symbolic operations that repress the objectionable idea, namely that the heroine, far from being the innocent maiden that she is portrayed to be on the screen, is in reality a temptress and a sorceress. In English, when a woman is regarded as only a sexual object, she is said to be a *fuck* or a *lay,* who is *taken to bed*, and the word *fuck* itself in polite society is sometimes called the "F-word." Thus, the name Morgan le Fay is a perfect crystallization of these ideas because there is an *F* "embedded" in a *lay*, namely *Le(F)ay.* If the F element is removed, the remainder is still *lay*, which retains the original idea, so the content is further disguised by rearranging the letters and substituting *i* for *y: Leay* → *Leya* → *Leia.* As for Morgana, the capital letter is removed, or, more precisely, the word itself is de-"capitated," giving the transformation *Morgana* → *M* + *organa.* Also, the *organa* element is then capitalized, making it the new capital or surname. Finally, the desexualized woman (Leia) is merged with the decapitated product of her own sorcery (Organa) to give us the insipid image of the Princess herself, dressed in a white smock like a resurrected virgin, with her hair done up in the style of a Hopi Indian maiden.

In the *Star Wars* trilogy, each intermediate stage of derivation in the deep structure generates its own surface structure imagery. This process is shown in Figure 3.3, in which the *lay* of Morgan le Fay is combined with the *fata* left over from Fata Morgana to become the Jabba the Hutt scene in *Return of the Jedi.* Described in one of the spinoff books as a wicked, repulsive blob of fat, Jabba the Hutt is shown sucking on a hooka pipe, like the caterpillar in *Alice in Wonderland.*[9] Visually, Jabba has a broad, toad-like mouth that dominates his entire face, and this orality is reflected in his name, which is composed of a pair of "fleshy" bilabial stop consonants (b...b) and a pair of mammary t...t's. This is no coincidence, for Jabba's henchman, Boba Fett, has a name that exemplifies the same principles (bb...tt), as does his lieutenant Bib Fortuna (bb). Significantly, this is the only scene in the trilogy in which we see Leia's body: she is chained to Jabba's throne wearing a bikini,[10] each breast accentuated by a coiled snake of wrought gold. Her hair is braided into two intertwined strands that hang between her breasts. Moreover, Leia shares the frame with Jabba's prominent hooka pipe, a pair of hemispheric vessels as large as she is. Leia's bare skin is enhanced by Jabba's throne, which is itself a pile of animal skins, and by the scalps hanging from the bounty hunter's belt. Thus, the Jabba the Hutt scene is a pastiche of the elements that have been removed from the female prototype to create Leia the white-robed princess: the body fat, the nurturing breasts, the snake-like tresses, the hairy patches of skin, and the viper-like lures of her sexuality. With this constellation of imagery, can the vagina dentata be far behind? As if on cue, Jabba decrees that the male heroes are to be thrown into a pit inhabited by a monster, the Sarlacc of Carkoon, which is a huge slimy mouth surrounded by teeth.

In *Star Wars*, not only in this example, but throughout the entire trilogy, there is a schism between the surface structure commentary on events and the imagery of the events themselves. These movies operate on two distinct planes: on one level, they are adventure stories about good versus evil and freedom versus tyranny, but at the same time they reference traditional mythological elements in such a way that a native speaker of English will subconsciously make a very different interpretation of the images on the screen. Leia Organa, who is visually portrayed as a lily-white maiden of exemplary virtue, has a name that is derived from Morgan le Fay, and in the Jabba the Hutt scene, the seductive aspects of her body and its consequences for manly virtue are unequivocally displayed. The negative connotations of Leia's character rationalize the actions that are done to her, transforming her in the audience's eyes from innocent victim to deserving recipient. In the first film alone, Leia is stunned by a laser gun, shot with a hypodermic syringe, forcibly shown the power of the Death Star (which vaporizes the planet Alderon), and confined to a prison cell. When she is finally rescued by

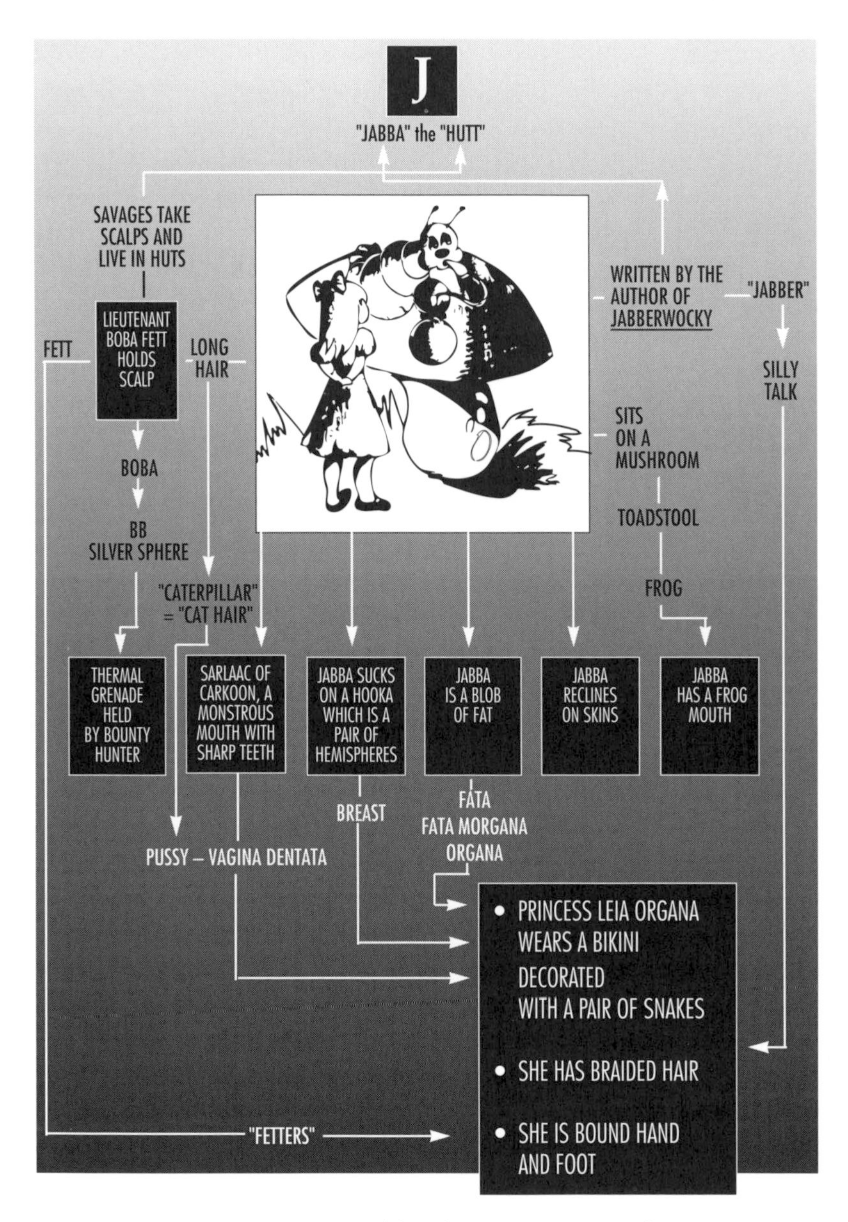

Figure 3.3 *Star Wars'* Jabba the Hutt is a transformation of the caterpillar scene in *Alice in Wonderland.*

the heroes, she expresses her thanks by giving them orders in a haughty tone of voice, insults Han Solo's sidekick, and then leads them all into a foul-smelling garbage chute, which drops them into a cesspool inhabited by a slimy monster. So much for female gratitude and leadership.

In the *Star Wars* series, there is a profound discrepancy between what the movie *says* it is doing and what the images actually convey. It *says* it is portraying women as leaders by making Princess Leia head of state and the matronly Mon Mothma (who?) commander of the rebel forces when in fact the real heroes are Darth Vader, Luke Skywalker, and Han Solo. By filling the screen with alien creatures bustling about, the movie *says* that the galaxy is a polyglot, multicultural society, but in fact all the nonhumans in the films, with the exception of Yoda, are either subordinate to humans or only nominally in charge. The characters say that the Force is a spiritual power, but in fact it always manifests itself as military superiority. The titles say that the Empire is evil and the rebels good, but there is little difference between them in their behavior or concerns, which invariably involve weapons development and plans of attack. In short, the *Star Wars* movies exemplify an Orwellian use of language and imagery, in which Sexism is Equality, War is Freedom, Planets are Targets, and all moral distinctions are collapsed into the martial and mechanistic vision of the space-race technology itself—just like the society that produced it.

Dark Invader

Nowhere is the discrepancy between deep structure and surface structure more apparent than in the dramatic weight of the "hero" relative to that of the "villain." The ostensible hero of the series, Luke Skywalker, is a boring nonentity, but Darth Vader is a memorable creation of the film maker's art. Who *is* this masked man? In our deep structure model, the anagrammar of Anakin, D.V. is doubtlessly V.D., venereal disease; and the history of V.D. in the past decade is largely the history of AIDS.[11] I realize, of course, that AIDS was not yet "discovered" in 1977 when the first movie was released, but, like moviegoers, let us suspend our disbelief and follow these characters through to the end. Let us assume that the dark invader is the immune deficiency syndrome, or IDS, nowadays known as the HIV virus or AIDS. In support of this interpretation, the term HIV contains *IV* as does the title *Episode IV*, and, second, the word *AIDS* contains *ids*, which alludes to Freud, the doctor of sexual repression, so it is highly appropriate symbolically. Also, one gets AIDS by exchanging body fluids, particularly by sexual intercourse, but also from IVs, injections, and contaminated blood; and we have already noted the black robotic sphere that inoculates Princess Leia. What does the HIV virus look like? Well, the published photomicrographs show it as a sphere covered with short appendages, like a thistle with blunt thorns—not exactly a carbon copy of Darth Vader's automated syringe but nonetheless a sphere with inoculative powers. Did George Lucas know about AIDS as early as 1977, even earlier when production time is taken into account? Probably not, but he did know about the values of the technocracy, its images, and aspirations; and he understood on a symbolic level that if there were no AIDS it would be necessary to invent it.

AIDS is not just a disease; it is also a symbol—retribution for illicit sex and drug use, and this complex of ideas has a long history in American culture. The anthropologist Emily Martin, who has studied the symbolic content of modern medical literature, concludes that Americans see the human body as a culturally constructed barrier between themselves and outside invaders.[12] This barrier is also thought of as a machine that wages war on alien microbes, which must be recognized, neutralized, and ultimately exterminated. The medical expression of this symbolic complex is the immune system, a cluster of physiological responses, including white blood cells, antibodies, and cellular memory, which is specialized for recognizing and eliminating intruders into the body. Since the physiological immune system is the culture's first line of defense against alien invaders, a disease like AIDS (acquired immune deficiency syndrome), which is a permanent breach in the biological barrier, is particularly terrifying, for it leaves the body exposed to all other sources of infection. Moreover, the way in which AIDS is transmitted potentiates these fears, for it feeds into the puritanical distrust of the body as a source of sin, seems to confirm the idea of physical retribution for immoral acts, underscores the subversive role of homosexuals and drug addicts, and highlights the danger of pollution from such bodily fluids as blood and semen.[13]

In conformity with the imagery that underlies AIDS, the *Star Wars* trilogy presents an endless succession of barriers being breached. The first film begins with Darth Vader's Imperial storm troopers bursting through the defenses of Princess Leia's ship and taking her prisoner. Then the bad guys destroy Luke Skywalker's home and kill his foster parents. When the heroes' ship is pulled into the Death Star by a tractor beam, the heroes, disguised as storm troopers, break through the Death Star's defenses, rescuing the princess from her cell. In the rebel counterattack, the somatic imagery is made even more explicit, as the heroes fly their ships up "the thermal exhaust port, right below the main exhaust," which is the "only vulnerable target" on the Death Star. In *Return of the Jedi*, the plot hinges on destroying the power source of the Death Star's invisible deflector shield, which makes it otherwise "impenetrable."

Thus, *Star Wars* shares with AIDS *the symbolic concept of the body as a machine that protects against alien invaders that penetrate it during acts of sexual intimacy or through contact with contaminated blood*, but the movie focuses on the overt body of defense and counterattack, while the underlying motivation of contaminated body fluids is barely hinted at in the surface structure. Nonetheless, it is the imagery of infection through sexual contact that ties together the sorceress Morgan le Fay, the desexualized Princess Leia, the scene with Jabba the Hutt, and the references to V.D., IV, and dark invaders. In fact, the very concept of the hero in *Star Wars* is almost coextensive with *fluid-free bodies* constructed of mechanically generated parts. Lightsabers are the weapons of choice for Jedi knights because they cut clean, presumably cauterizing the wounds they make. When Darth Vader chops off Luke Skywalker's right hand with a lightsaber, not a drop of blood is visible on the screen, and no body fluids of any kind appear in any of the *Star Wars* movies, even though the heroes shoot their way through throngs of Imperial storm troopers.

In *Stars Wars*, flesh and blood are conceptualized as primitive, whereas heroes are made of radiant energy. Princess Leia, Luke Skywalker, and Ben (Obi-Wan) Kenobi make appearances as lifelike holograms, which in the case of Leia and Luke are projected from the head of R2-D2. In other words, these characters are unequivocally good because their biological bodies have been encoded in mechanical brains and regenerated as waves of light. In the same vein, Jedi warriors fight with laser-like beams of light portrayed as hand-held sabers; and in the case of the Emperor, who is even more closely connected to the Force, albeit to its Dark Side, crackling, blue, light waves of energy emanate directly from his hands. The equation of spirituality with light and baseness with fluids is also conveyed in the scenes with Yoda, the alien trainer of Jedi knights. In this sequence, Luke and the faithful R2-D2 crash-land in a marsh on Dagobah, which is entangled with vines and inhabited by ominous aquatic creatures. There Luke meets Yoda, who begins to train him in levitation, psychokinesis, and other "spiritual" skills. These so-called spiritual strengths are themselves emblems of disembodied action, and the diminutive Yoda demonstrates the Force by lifting Luke's spaceship out of the primal marsh by the power of mental concentration alone. Significantly, a few scenes later, Han Solo mentions matter-of-factly that the Imperial fleet always dumps its garbage before going to light speed.

The Y-forked Organ

The contrast between contaminating fluids, associated with biological processes, and a spiritual power, manifested as mental control of radiant energy, creates a corresponding bifurcation of the human body into those parts that are associated with pollution and those associated with intellect and the perception of light. In the technocratic world view, where heads are considered "advanced" structures, the hind parts are necessarily "primitive" and often threatening. The *Star Wars* movies exemplify these assumptions in a variety of ways. First of all, evil characters typically are masked, so that their human facial features are disguised. The Imperial storm troopers wear face protectors and full body armor; the bounty hunters wear ominous helmets; and Darth Vader himself is both helmeted and masked. At the court of Jabba the Hutt, almost everyone is masked except the heroes and the alien species. The good guys, in contrast, never wear armor except as a disguise, and even when they don aviator helmets to go into battle, their faces are left exposed. Almost the only exceptions to this rule are the Emperor, who wears a monk-like cowl, and his officers and crew, who wear uniforms like those of the German army. Alien species, of course, are exempt from this rule because it is their odd or deformed heads that mark them as alien in the first place: Jabba, Yoda, Akbar, and so on.

But if the Emperor and his crew are evil, why are they portrayed with human heads? The Emperor is said to embody the Dark Side of the Force, the same Force that throbs in Luke Skywalker, and we are told that even Jedi knights can be captured by the Dark Side, as was Luke's own father Darth Vader. So the evil Empire and its rebels are really kinfolk after all, as closely related as father and son, so it is hardly surprising that both have human forms. Notice, however, that this kinship system is rigorously patrilineal, for both Luke and Leia admit to having no recollection of their mother, and nowhere does she appear in the plot. Although Leia is said to be Luke's twin sister, the closest thing there is to motherhood in the entire trilogy is a stern, white-robed woman, the putative commander of the rebel alliance: Mon Mothma, a series of nasalized bilabials (m...m...m) in a war room full of men. In the technocracy, heroes are born of light and not of woman.

This suggests that the head is not only the most advanced part of the human body but the most *male* part as well. In Nebuchadnezzar's Dream, for example, it is no accident that the golden head of the statue is that of a king, not a queen, and that the most fragile and ignoble element, namely clay, is the material which partly forms the feet. Moreover, these anatomical metaphors are endemic in the English language. In American English, for example, it is common to refer to people by the body parts that exemplify their social roles, such that sex objects can be denoted as *tail* or *ass*, obnoxious persons as *assholes*, and ethnic groups by idealized skin color, such as *Whites* and *Negroes*. Also, as the anthropologist Edmund Leach points out, the equation

deity = human phallus = totemic animal = human head =nose

probably occurs in all known human mythologies. Moreover, the totem animal itself is often a species with prominent headgear, such as beards and dewlaps, plumes and horns, antlers and cockscombs, and this principle is exemplified in English by the word *stag*. English speakers who have never even seen one of these animals in their lives and who may even think it a distinct species know very well that it means "lone male" or "men only," as in *to go stag* and *stag party*. Linguistic techniques show that the word *stag* is vintage English, traceable through Middle English and Anglo-Saxon right back to Proto-Indo-European, the ancestral language from which most European tongues diverged. The Proto-Indo-European root of *stag* means "to sting" or "to prick." *Prick?* This loanword from German has become American slang for "penis," so even in ordinary English we can see the symbolic equivalence of the totem animal, the phallus, and the human head as pointed out by Edmund Leach. Moreover, in imagery the world over, the head of the body is associated with the dominant power relations in a community, as with *heads of families* and *heads of state*.

But do not women have heads too? Anatomically perhaps, but in the androcentric mythology of the technocracy, females have only *one* head whereas men have *two,* connected to each other by a wooden trunk to form a Y-shaped architecture. The *upper fork* is the male head in social and intellectual aspect, the *lower fork* the energetic, phallic aspect. In the war-room scene of *The Return of the Jedi*, where the rebels plan their attack on the second Death Star, all three kinds of human heads are on display together: desexualized female heads in the form of Leia and Mon Mothma; powerful male heads in the form of Luke, Han Solo, and the soldiers; and a third head, the phallic head, in the form of nameless, funny-looking creatures with enormous hairless heads covered with shiny skin and punctuated by a pair of thick, fleshy eyestalks, like the eyes of snails and slugs. In this androcentric system, where the organ of procreation is the head, the phallus itself is accorded cephalic properties, such as eyes, brain, and the power of independent action; and this idea, while seemingly idiosyncratic, is in fact one of the most venerable concepts in the human mythological repertory. The Y-forked concept of the human head, combining both the phallus and the animal head into a single composite image, is one of the first ideas ever to be represented in symbolic media by the human species. From the Upper Paleolithic in France, the period with the oldest artistic works known to archaeology, excavators in the last century found a Y-forked piece of reindeer antler in which the prongs are carved into both a human phallus and the head of a snake, which dates from about 30,000 years ago.[14] Also, in the mural art of that era, there are animal heads topping human figures with erect phalluses; and the male head, divorced from the rest of the body, occurs as a design motif, as at the rock shelter at La Marche, where dozens of examples have been found. Thus, the iconic deep structure of the technocracy, with its disembodied male heads and Y-forked phallic/cephalic structures, is perfectly consistent with both the comparative study of human mythology and the archaeological history of imagery.

The imagery of the Y-shaped male organ is almost as important to technocratic mythology as Nebuchadnezzar's Dream, for it encodes the masculine properties associated with the golden head of the king, namely, phallic and intellectual power. In science, the life force in male aspect is represented by the Y chromosome, but in *Star Wars* it is conveyed by the scenes with Yoda, a shriveled 900-year-old alien guru on the swamp planet Dagobah, a world with "massive life-form readings." In Lucas's films, the hot spots of the technocratic mythology are realized by visually powerful images, such as Darth Vader, that strongly contrast to the stunning two-dimensionality of a Leia or a Luke; and Yoda is an appealing character, whose look and set was closely supervised by Lucas himself. In fact, the director of *The Empire Strikes Back*, Irvin Kershner, says that the Yoda sequence is "the heart of the picture."[15] Significantly, this little, misshapen creature who embodies the Force is almost all Y-shaped head, with enormous fleshy ears pointing out to the sides, and even his name begins with a "Y." Timelessness is conveyed not only by Yoda's great age but also by his name, an anagram of *today,* and by his large, pointed ears, for *Y* + *ears* = *Years*. Also, as if to emphasize that this particular character inverts the symbolic order of the two male heads, giving primacy to phallic instead of intellectual power, Yoda speaks in portentous subject/predicate inversions, such as "Fear you will know soon." Appropriate too is that Yoda's voice is played by an impersonator with the name of Frank Oz—reminiscent of the disembodied head of the wizard.

In *Star Wars*, the shaft of the "Y" is also a literal character. In the androcentric mythology, the shaft that connects the two forks in the man's body is not made out of flesh and blood but *heartwood,* a bloodless, generative substance that contrasts to the bleeding hearts of wimpy males; and this image is embodied in the Wookiee Chewbacca. This alien creature, covered with long, shaggy hair, is Han Solo's ever-present sidekick, and he seems to violate all of the rules of gender-appropriate behavior. He is a nonhuman creature, not a machine, yet he is perennially associated with the real heroes of the story, accompanying them on their adventures. Even though he is covered with hair, he is both a fierce fighter and a skilled pilot, and in no way effeminate. Although intelligent enough to play three-dimensional chess and to reassemble the disarticulated C-3PO, he is nonetheless completely inarticulate, communicating only in emotional screeches. Also, he not only appears bigger than the other characters; he is in fact played by an actor who is over seven feet tall. And what are we to make of his odd name?

Although chewing tobacco is the most masculine of cuds, the vegetative associations of the Wookiee have far deeper roots than this. We first meet Chew*bacca* in a bar, which is a place that sells fermented beverages; and historically in European culture, fermentation is associated with the wine god Bacchus, from *baccate,* which means "bearing berries." There is also a homonymous word, *baccalaureate,* which is cognate with *baculum* or staff, such as what a shepherd carries, and it is semantically associated with the "sheepskin," which corresponds to the Wookiee's hairy pelt. Princess Leia even calls Chewbacca a "walking carpet." In addition, there are strong chess associations. *Wook* sounds like *rook,* which is a chess piece, one of a pair of crenelated towers, also known as castles, that typically protect the king in the end game. That the Wookiee is a chess piece as well as a berry is shown by the fact that he plays chess and sometimes wears a bandolier of alternating black and white squares. Also, Han Solo merges the two images when he says that when Wookiees lose at chess they like to tear the arms off their opponents—the image of a trunk without limbs. Thus, it is hinted at in the surface structure that Wookiees make trees, as do nuts and berries. Also, chewing tobacco, the surface etymology, is also a kind of leaf, reinforcing these vegetative associations. Moreover, in preindustrial European mythology, the equation of hair with leaves and blood with sap is commonplace. Thus, Chewbacca is the vertical shaft of the letter "Y," as solid as an oak and as germinative as an acorn.

When the phallic tine of the male fork is contrasted with the cephalic tine, the latter outranks the former, creating the bifurcation into higher and lower forks; but when the entire male body is contrasted with that of females, both male heads radiate with a captivating light. The pair of male heads in radiant aspect corresponds to the pair of male twins hatched from the same egg, the Gemini of the Romans and the Dioscuri of the Greeks. In the English language, the equation between male heads and radiant celestial bodies is further reinforced by the pun of *son* and *sun,* which Shakespeare used to good effect in the opening speech of *Richard III* ("Now is the winter of our discontent Made glorious summer by this sun of York."). Indeed, when we first meet Luke Skywalker, he lives on a planet with a pair of suns visible in the sky, one bright, the other clouded over. In *Star Wars*, the mythological twins (who are never completely symmetrical because one is left and one is right) are Luke Skywalker and Captain Han Solo, who often fight as partners against the Evil Empire, but who also go their separate ways when Luke is dealing with the Force. Han Solo is the less spiritual of the two, with the attributes of an ancient sun god (*Solo* as in *Ol' Sol*): like the Egyptian Ra he flies in a ship called the *Falcon,* keeps Luke from freezing to death on Hoth (Thoth, the Egyptian moon in male aspect), is tortured by "clouds" (that is, by the ruler of the Cloud City of Bespin), can be frozen by carbonite (that is, "black night"), moves in concert with the moon (that is, with Leia on the Moon of Endor), and is consistently associated with Chewbacca, the male vegetative force. Where Han is a good right han(d) man, Luke's right hand was cut off by Darth Vader in Episode V and replaced by a robotic device, suggesting that his nature is further to the left, more spiritual and cerebral—which in the technocracy is equivalent to cephalic control of radiant energy.

Luke's spiritual mission is reflected in the name *Luke,* which is probably short for *Lucifer*, "the light-bearer," and an eponym of Lucas himself. As George Lucas tells Alan Arnold, "I wanted a concept of religion based on the premise that there is a God and that there is good and evil.

I began to distill the essence of all religions into what I thought was the basic idea common to all religions and common to primitive thinking."[16] Clearly, the name *skywalker* is not what it seems to be, for "one who walks in the sky" is still too down-to-earth for anyone who can distill the essence of all religions into three adventure films. If Princess Organa is any indication, then the last name of the character is closer to the deep structure derivation than the first name, and this suggests that the prototype of *sky*walker is not the obvious surface structure "sky" at all but the Proto-Indo-European root **skei-*, "to cut" and "to split"—the origin of the word *science* itself, the real religion in the series. The root **skei-* is the ancestor not only of the Latin *scientia* but Celtic "knife," Greek *schism,* and English *shit.* Thus **skei-* is the process that separates the pure from the polluted, the orthodox from the heterodox, the enlightened from the ignorant. Luke Skei-walker is **skei-* in the aspect of science, whereas the evil Anakin Skei-walker (Darth Vader) is **skei-* in the aspect of male shit (Lord of the Sith). These two aspects of **skei-* are "split" even before Luke is born by the affinity of Anakin for the Dark Side of the Force, which even the enlightened Ben Kenobi is powerless to prevent. In the same way, the mother of the celestial twins, Luke and Leia, disappears at birth, split off, apparently forever, from her offspring. Next, the male and female pair, Luke and Leia, are themselves parted at birth by Ben Kenobi in order to protect them from the Empire, that is, from the Dark Side. Finally, the attributes of the temptress and sorceress Morgan le Fay are split from the white-robed Princess Leia, where they pool in the throne room of Jabba the Hutt. *To cut, to chop, to split, to separate:* This is indeed the essence of the technocratic religion.

Splitting the body into mirror images helps to sustain the one-two punch. First, by defining one part of the body as alien nature polluting to the purified remainder, it necessitates a process of ritual purification that cuts off, symbolically or literally, the offending organ; and, second, it creates a new, dark force, cut off from the body, that threatens to re-enter and pollute it. Thus, purity and pollution form a single, self-sustaining cycle, which is why Luke Skywalker and his companions must penetrate the Death Star and remove the purified female element (Leia) before they can destroy the Death Star. The incorporation of purity and pollution into a single loop also explains why Sky(**skei-* ,"to cut") is paired with Walker, an apparent contradiction because a walker is in contact with the ground. But in the movie, the contradiction is resolved by the imagery of what are called "Imperial walkers"—huge, quadrupedal machines that attack the rebels on the ice planet Hoth. The name *walker* (pronounced in English with a silent "L") is cognate with the Proto-Indo-European **wak-* or "cow," and with Spanish *vaca* ("cow") and *vaquero* ("cowboy"). The cow is historically the source of the first vaccine, the smallpox vaccine; and the root of *vaca* is the linguistic source of the English word *vaccine*. Thus, the name Skywalker is a condensed image of the technocratic process, encoding both nature as a source of pollution (**skei-*, "to split off," "shit") and as a source of power if purified (**wak-*, "cow," "vaccine"). The full name, *Luke Skywalker, Son of Anakin*, encodes the entire technocratic mythology: namely, "the light-bearer who parts the cow shit (of nature)—in order to distill the pure inoculate that protects against pollution." In science as in Nebuchadnezzar's Dream, feet are soiled but heads are golden.

The Logic of Pure and Polluted

The *Star Wars* trilogy encodes the man-making process that produces the Y-forked organ and the golden head of Nebuchadnezzar. This process requires that radiant energy, epitomized by celestial light, be separated from fluids, especially blood. That is, women are born, but men are made by splitting themselves off from women and nature through ritual acts that either "shed blood" (their own) or show them to be literally "brilliant." These ritual acts of blood and fire, by separating the male and the female elements, leave behind bloody stumps and pools of body fluids that are by definition female and polluting.

The only relatively non-polluting body fluids in this system are *white* fluids, namely semen and mother's milk; but even these are gender-sensitive, for the color white is good in the case of females, as with the white-robed Princess Leia, but bad in the case of males, as with the white-armored storm troopers. Semen is a biological substance produced by the lower parts of the body in proximity to the organs of elimination, a fact which no amount of mystification can completely obliterate. For this reason, semen, although an unequivocally *male* fluid, homologized to stardust, is still a polluting substance; and the penis, although defined as a head, is conceptualized as the *lower* fork of the Y, inferior to the head of radiant light. Mothers' milk, the other white bodily fluid, is not high status either. In fact, in the United States it is generally considered bad even for mothers and infants, for the latter typically are fed store-bought formulas from sterilized bottles. [17]

In *Star Wars* this difference between head and phallus is expressed by the split between relatively *terrestrial* males such as Han Solo, who has a history of the illicit crossing of boundaries (both smuggling and snuggling—he kisses Princess Leia), and full-fledged *celestial* males, such as Luke Skywalker, Darth Vader, the Emperor, and Obi-Wan Kenobi, who live in monastic isolation from women; wear black, monk-like cowls; and have apparently completely transcended a biological level of function. In this so-called "spiritual" realm, even semen is polluting; and Jabba the Hutt is portrayed as the embodiment of evil because he *likes* to suck fluids, presumably tit (the double-hemisphere hookah).

In the man-making process, the splitting of males from females occurs at birth, when the male baby is forever separated from his mother (female babies linger a little longer, which explains why Leia remembers her mother's smile but Luke remembers nothing at all). Moreover, the twin male and female in *Star Wars* suggests an androgynous infant that has been split into male and female "halves" by an act of purification, such as circumcision of the infant shortly after birth. In the United States, until recently, almost all male infants were circumcised, irrespective of the religion of their parents, which is very different from Europe, where this practice is typically restricted to Jews.[18] In American popular culture, circumcision is defended in terms of cleanliness and reduced likelihood of infection, a rationale that attests to its connection with the logic of purity and pollution. Although the practice does have a biblical charter, the New Testament declares it irrelevant to Gentiles, but for Jews it is still a sign of the covenant between God and his chosen people. In *Star Wars*, there are no direct references to circumcision, but in *The Return of the Jedi*, the skin-headed alien in the war-room scene, which in my interpretation represents the phallus, is named Admiral *Ackbar,* an Arabic word meaning "great" and an appellation of Allah. This character is probably a veiled allusion to the custom of circumcision, which is associated with Jews and Moslems in American culture, and appears as a covert theme in *Star Wars*. Luke's benefactor, Ben (Obi-Wan) Kenobi, has a first name that is Hebrew for "son of"; and it is Ben who separates Luke and Leia at birth. Thus, the spirituality of males is further indicated by the act of circumcision, which literally "cuts off" (a piece of) the male organ from the rest of the body and marks the boy as one of the chosen people of God. Unlike Islam or Judaism, however, American culture views the act of circumcision as a routine medical procedure, done while the infant is still in the maternity ward, and there are no rituals surrounding either the act itself or the disposal of the prepuce, so it does not in itself "make men." That comes in adolescence, when the young man is cut off from his home and incorporated into rituals of blood and fire, of which there are two distinct types, the military and the scientific.

Both of these man-making processes—the military and the scientific—are combined in the character of Luke Skywalker. At the beginning of the series, the adolescent Luke is split off from his home when his foster parents are literally killed by the soldiers of the Evil Empire, catapulting him into a world of blood (not shown) and fire (abundant) which he is powerless to fight. At this point, the retired trainer, Ben (Obi-Wan) Kenobi, takes Luke's career in hand and helps him to become a Jedi knight, an ancient class of warriors who defended the republic against the Empire and whose strength is based on the mental control of the Force. Thus, the plot of the Star Wars series is essentially the transformation of boys into men, which is to say, the differentiation of "male" at the expense of "female" parts of the body. Like all paranoid fantasies, the purity and pollution cult of industrial civilization is an exquisitely logical process that can, so to speak, be flow-charted (Figure 3.4).

Although, as *Star Wars* confirms, both science and war have the same ultimate goal of transforming boys into men, and they can be embodied in the same person, the two institutions are nonetheless distinguished by their very different strategies for dealing with female pollution. In the military strategy (or in its civilian analog, gang violence), the young man is trained to "shed blood"—namely that portion of his *own* blood that has been contaminated by women. This contaminated blood is symbolically equated with the dark, deoxygenated blood returning to the heart, which is also known as venous (that is, "Venus") blood because of its female associations. Male blood, however, is bright red blood, equated with arterial circulation, which is the sign of a strong, healthy heart, the seat of courage. This imagery is chartered in the mythology of industrialism, for who does not know that the first act of modern scientific medicine was Harvey's discovery of the circulation of blood?

When the imagery of male blood is assimilated to the logic of purity and pollution, warrior males can shed venous blood by shedding the blood of other males. Shedding female blood, or one's own blood in isolation, is useless in this respect, for it only contaminates one further. Thus, it is not blood per se but the social relations of male solidarity and male conflict that are the real purifying agents; and for this reason, in *Star Wars* male "blood shed" is depicted as a social activity, either

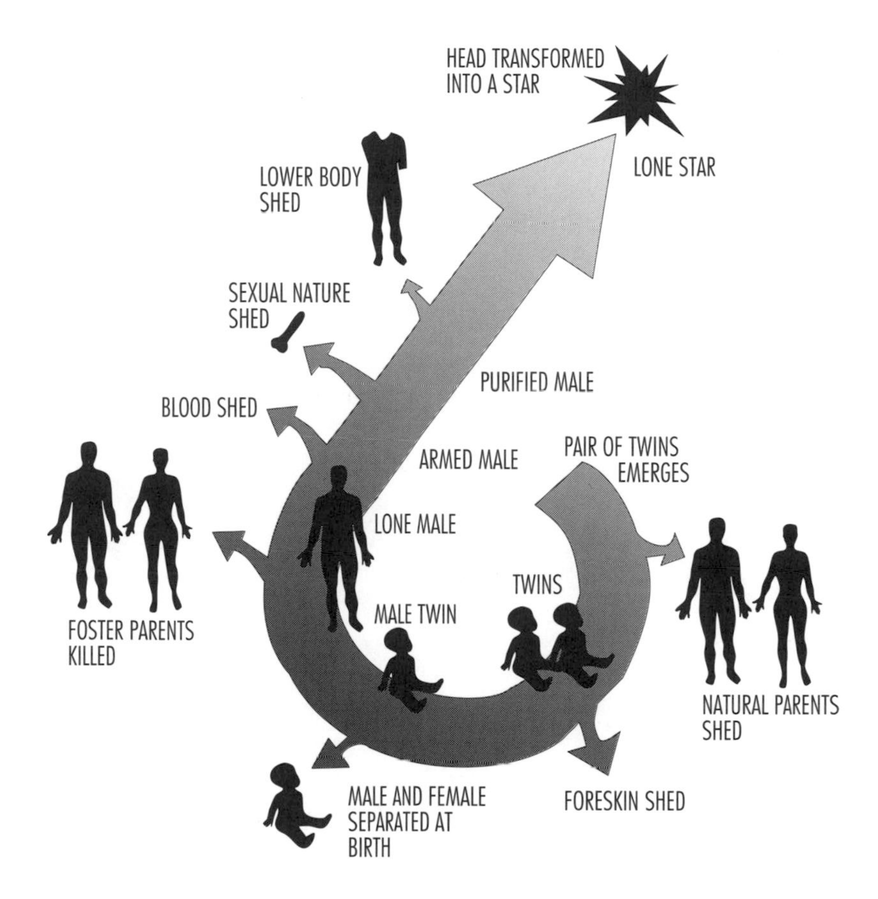

Figure 3.4 To purify the human body, first shed one's blood, then transform the remnant by fire.

the confrontation between armies, as in the concerted attacks on the Death Stars, or as one-on-one confrontations between warriors, as in the fencing scenes between Luke and Darth Vader.

In *Star Wars*, however, "blood shed" is conceptual, not biological, for men are made of light, not fluid; and attack is most effectively implemented by phallic- or cephalic-shaped objects that symbolize the male organ and simulate its association with celestial fire, such as guns, bombs, rockets, and other spherical or cylindrical devices that create explosions of light from a point source. The lightsaber, the weapon of choice for Jedi knights, also combines both the phallic shape and the imagery of celestial fire—but unlike the weapons of the purely military heroes in the movie, it appears to be bloodless in its effects, killing without spilling. This weapon is more consistent with the scientific approach to purification, which is based not on the shedding of blood, but on making the cephalic fork more "brilliant" through the mental control of female nature.

The scientific pathway of purification (the upper fork of the "Y") is represented by the spiritual side of Jedi training. The name of Luke's benefactor, Ben (Obi-Wan) Kenobi, contains the Scottish dialect term for "to know" (*ken*), as well as the anagram of the English word *know*. In addition to his general knowledge, Obi-Wan is in contact with occult forces, and his name is suggestive of *obeah,* a form of black magic practiced in Guiana, a tradition probably known to Lucas through his interest in comparative mythology. In one scene, Ben (Obi-Wan) Kenobi, the trainer of Darth Vader and the benefactor of Luke, wears a *kimono* with a sash tied around the waist, reminiscent of a samurai warrior, and the Japanese word for "sash" is *obi*. The aristocratic element is present too, for *O.B.E.* means Order of the British Empire, a rank of knighthood; and the part of Kenobi is played by a knight in real life, Sir Alec Guinness. Also, Yoda, the other professional trainer of Jedi knights, has a strong propensity for Zen-speak, which any Californian knows is the religion of the samurai and the spirituality of choice for contemporary technocrats. Thus, in *Star Wars*, the "learned" characters are warriors and warlords, whose spirituality is measured by their prowess with a lightsaber and their willingness to use it. This portrayal of technocratic enlightenment is exactly right, for in natural science, as well as among modern academics generally, the search for truth is a search-and-destroy operation in which one "explodes" previous theories by a "brilliant" piece of work that establishes the scholar as the "top man" in his field—a *battle-field* presumably—while reason itself has become indistinguishable from sexual assault, full of probing questions and penetrating arguments. In *Star Wars*, as in the technocracy itself, knowledge and warfare are viewed as two aspects of the same system, differentiated primarily by their strategies of purification: warfare sheds contaminated blood through rites of male solidarity, whereas science polishes the cephalic head, making it more brilliant. Thus, the modern technocracy is an inherently mythological enterprise, a system of symbolic transformation processes that can be summarized as follows:

1. The first six axioms (including this one) are taboo, and may not be mentioned in writing or formal discourse.

2. The technocracy is the transformation and purification of "female" nature by means of celestial light and a "male" apparatus.
3. A "male" apparatus is one that mimics the imagery of the *Y-forked organ*, namely a Y-shaped piece of heartwood that branches into a phallus with eyes and a brilliant human head. The Y-forked organ and its transformed products have the additional attributes of *celestial, pure,* and *fiery.*
4. Non-celestial light (that is, light reflected from terrestrial, lunar, or female objects or surfaces) acquires "male" attributes when transformed with a Y-forked organ (the role of photography and related imaging processes).
5. By participation in the technocracy, the biological body of the male becomes a disembodied Y-forked organ.
6. The disembodied Y-forked organ leaves behind a "female body," that is, a biological residue consisting of lower limbs, trunk, gullet, and female genitalia. Femaleness, which is inferior to maleness, entails the additional contextually sensitive attributes of *terrestrial, lunar, incomplete, bloody,* and *contaminated.*
7. When eNobeled by the King of Sweden or by a comparable agency, the disembodied Y-forked organ is transformed into the symbolic equivalent of a celestial body—that is to say, a *star.*
8. The phallic fork of the Y-forked organ dies, but the eNobeled head is physically immortal, the genitor of a new species of man and of a more advanced society. (This is Nebuchadnezzar's Dream.)

The mythology of bloody stumps and exploding stars could easily be dismissed as absurd psychologizing were it not for the fact that the political life of Europe has been convulsed within living memory by purity and pollution cults very similar to this. As the Austrian scholar Klaus Theweleit has shown in a two-volume study of fascism, *Male Fantasies*, based on analysis of the books and images produced by members of the Nazi Party, the Freikorps, and other militarist reform movements in the German-speaking countries between the World Wars, these groups of soldier males (as he calls them) were held together by androcentric mythologies that celebrated male solidarity and unity, which in turn was symbolized by the male head in both phallic and martial aspect, and expressed through acts of violence that had the intention of reducing one's male opponent to the state of "a bloody pulp"—which was also their image of women. As Theweleit expresses it: "The texts of the soldier males perpetually revolve around the same central axes: the communality of male society, nonfemale creation, rebirth, the rise upward to hardness and tension; the phallus climbing to a higher level. The man is released from a world that is rotten and sinking (from the morass of femaleness); he finally dissolves in battle."[19]

Similarly, Arno J. Mayer, a professor of history at Brandeis University, has analyzed the roots of the Holocaust in his book *Why Did the Heavens Not Darken?*, and he comes to the conclusion that the Nazi program was based on a comprehensive theory of blood and soil (*Blut und Boden*) in which so-called genetically inferior groups, including not just Jews but also Gypsies, criminals, and mental defectives, were seen as sources of pollution that caused diseases in the body politic. In fascist theory, Germany could be cured by restoring the body to its original wholeness (that is, by regaining lost territories) and by acts of purification that cleansed the "blood" (that is, the gene pool) by removing alien elements. This purification process entailed both the elimination of the impure elements and the regeneration of the pure male spirit through martial acts of bloodshed and fire. Significantly, the German term that is usually translated into English as "Master Race," namely *Herrenvolk,* means literally the race of "men" or "masters"; and Theweleit points out that it is necessarily composed only of males, for (in Nazi theory) men are the creative, intelligent, and culturally advanced element, whereas "Women ... are not even Aryan (though they become so if they give birth as 'white women' to a quantity of master-sons)."[20] In other words, the master race is the purified male body, freed of its polluting elements and cleansed by celestial fire. Thus, fascism, in its most virulent form, is primarily a cult of purity and pollution that separates the "male" body from contamination by women and other inferior groups.

Fifty years later, *Star Wars* is celebrating these same themes. In the trilogy, the driving force of history is a militarist male solidarity in which women, when not actual captives, are, at best, spectators of male violence, nominal participants, or desexualized sister-companions in the manner of Diana the Huntress. More significantly, woman's contribution to birth, growth, and nurturance is effectively denied, for the children have no memory of their mothers; their teachers are black-robed hermits or Y-shaped aliens; and only their fathers reckon in the genealogies. The beauty and sexuality of women is reduced to the court of Jabba the Hutt, in which the half-naked princess is exhibited as a captive, associated with vice and pollution, and paired with a monster vagina dentata.[21] Moreover, *Star Wars* sucks its energy from a perpetual state of war, with stages of history demarcated by attack and counterattack, victory and defeat, with no discernible difference between the good guys and the bad. Yet it does not even have the honesty of the Nazi, who admitted to loving war and hating inferior races, but wraps itself in allusions to freedom and democracy, to the defense of the republic, and to the rights of savages and women. Significantly, at the end of the trilogy, in the final encounter with Darth Vader, Luke—his transformation complete— is shown wearing knee-high, black leather boots and a close-fitting, black tunic that looks for all the world like an SS uniform without the insignia.

If *Star Wars* were just a movie, a simulation of events divorced from action, it would hardly be cause for concern, but on another level *Star Wars* is an artistic elaboration of the symbolic body of technocratic society, defining what a human being should be, the processes of individual development, the goals of technical activities, and the relationship of human society to the natural world. Moreover, as shown in the next chapter, the logic of purified bodies, as well as the imagery of bloody stumps and exploding stars, is definitive of the scientific method itself.

THE LONE GALILEO

Iconic Deep Structure and the Scientific Enterprise

Science isn't an alternative to folk taxonomy;
science is a folk taxonomy.

Myrdene Anderson[1]

To scientists, myths are false theories, a primitive stage in the development of scientific truth, but to anthropologists they are stories of past events that explain the nature of the world and justify community standards of collective action. All institutions have origin myths and culture heroes, and the enterprise of modern science is chartered by the tale of The Lone Galileo.[2] Like most legendary figures, Galileo is most famous for things he did not do. He probably never dropped an orange and a cannonball from the Tower of Pisa to show they fall at the same speed, for he understood the effects of air resistance and knew that they would not. He was not the first to use scientific instruments, for astronomical devices had been used for thousands of years. He was not the first to grind optical lenses, for eyeglasses had been available in Europe since the 13th century. He was not the first to apply mathematics to descriptions of nature—all the astrologers of the ancient world did that. He did not even invent the telescope, which was developed by anonymous contemporaries somewhere in the Low Countries. Rather, Galileo realized that the relation between theory and observation in physics is analogous to the logical deduction of theorems in pure mathematics. That is, theoretical propositions in physics correspond to axioms in mathematics, whereas observations correspond to theorems, so that the truth of a physical theory can be assessed by making the observations implied by the axioms. This is a very subtle idea—far too subtle to make someone a popular hero, so Galileo is remembered best for something that he did not do: the use of the telescope to prove that the earth goes around the sun.

A classic account of the origin myth of science is given in J. Bronowski's television series, *The Ascent of Man*, which presents Galileo as an unequivocally heroic figure: Galileo "was forty-five when he heard the news of the Flemish invention [the telescope], and it electrified him. He thought it out for himself in one night, and ... before he came to the Campanile in Venice, he stepped the magnification up to eight or ten, and then he had a real telescope."[3] As Galileo himself tells it in a letter dated 29 August 1609: "It is 6 days since I was called by the doge, to which I had to show it together with the entire Senate, to the infinite amazement of all; and there have been numerous gentlemen and senators, who, though old, have more than once scaled the stairs of the highest campaniles in Venice to observe at sea sails and vessels so far away that, coming under full sail to port, 2 hours or more were required before they could be seen without my spy glass."[4] In the classic accounts of the Galilean accomplishment, the invention of the telescope is a prelude to the birth of science. For example, in the words of Bronowski: "Galileo is the creator of the modern scientific method," and he did "for the first time what we think of as practical science: build the apparatus, do the experiment, publish the results."[5]

This portion of Galileo's life can be summarized as follows:

> *The Lone Galileo, Verse One:*
> A lone male genius
> Builds a tubular apparatus,
> In which lenses transform celestial light.
> He publicly displays the apparatus
> From the top of the Campanile,
> Amazes political and business leaders,
> And discovers aspects of nature hitherto unknown, namely:
>   Four moons of Jupiter,
>   The phases of Venus,
>   Sunspots,
>   Countless new stars in the Milky Way,
>   And mountains and craters on the moon.

The imagery of this myth exploits such culturally loaded terms as sun and moon, Venus and Jupiter, Venice and the Campanile; and even today, heavenly bodies have profound connotative meanings that extend far beyond their astronomical roles, conjuring up semantic associations to astrology, Greek mythology, romantic literature, and the history of science, to name only the most obvious. Also, there is the linear plot encoded in the verbs that resonates to Nebuchadnezzar's Dream: *builds* → *displays* → *discovers*. In addition, there is the imagery of the Campanile itself, the dominant architectural structure of Venice, and the symbol of a prosperous and sophisticated commercial society. The Lone Galileo draws on the power of its associated symbols to present a mythically potent vision of a technocratic world.

The Lone Male Apparatus

Like *Star Wars*, the charter myth of science encodes the imagery of the Y-forked organ. The very act of looking through a telescope merges both phallic and cephalic shapes into a composite image, such that, to a naive observer, the phallic-shaped object appears to emerge from one of the astronomer's eyes. Moreover, the phallic properties of telescopes are emphasized in the popular media by conflating the astronomer's inflexible tube with the telescoping variety favored by pirates, which can be shortened or extended at will. The phallic imagery of telescopes is also enhanced by the locations where telescopes are erected. When we first meet Galileo, he has mounted his tube on the top of a tower to look at ships at sea, and astronomical telescopes are still built in close proximity to the sky. In fact, the epochs in the history of astronomy are commonly demarcated by the size of the tools constructed on mountain tops: the 60-inch reflector on Mount Wilson (1908), the 100-inch reflector, also on Mount Wilson (1918), and the 200-inch reflector on Mount Palomar (1949). Nor has the advent of radio astronomy dampened enthusiasm for large optical instruments. A huge telescope of new design has been constructed on the top of Mauna Kea in Hawaii, and the 94-inch Hubble Space Telescope was launched by the space shuttle. Thus, the tubular shape of the optical telescope is essential to its role in the iconic deep structure of industrial mythology, whereas the newfangled radio telescopes, concave hemispheres better known as "dishes," are not symbolically equivalent at all.

Science, like other mythical systems, elaborates on the imagery of natural symbols, giving them its own particular interpretation, and for this reason the telescope is no ordinary phallus but a phallus with eyes. Physics recognizes the equivalence of lenses with eyes, calling the science optics, but the iconic deep structure takes this perception a step further, seeing lenses as extensions of the human body and not just as implementations of mathematical equations. The eye of the telescope is documented by the slang of astronomers. On Mount Palomar, the

Hale Telescope, a battleship-gray cylinder seven stories tall and housed in a dome the size of the Pantheon in Rome, is called the Big Eye by astronomers. A few hundred yards to the south, hidden in trees and underbrush, is a much smaller dome, housing an eighteen-inch Schmidt telescope, known as the Little Eye.[6] Also, it is not an accident that the astronomical eyes do not form an anatomical pair, side by side on the anterior end of the body, but are conceptualized as at lower and higher elevations, for the two eyes of the Galilean body are hierarchically ranked.

The prominent role of phallic imagery in dream and myth led Freud to the idea of the libido, which is a form of psychic energy based on the channeling of sexual motivation, but this psychoanalytic interpretation confounds the phallus as an organ of sex with the phallus as the symbol of male power. The ancient Greeks, for example, erected large stone phalluses, often surmounted by a male head, in front of their houses. These statues, called herms, did not symbolize sexuality per se but maleness in its social aspect, namely patrilineal continuity and domestic authority. In the same way, phallic imagery in science does not mean that scientific activity is necessarily motivated by sexual energy. Rather, the predominance of the phallus in scientific imagery has much in common with that of the ancient Greeks, namely, the use of a male body part to symbolize specifically male contributions to the social and natural order.

In science as in *Star Wars*, the status of maleness is achieved through a succession of cuts that separate the so-called male body parts from their prescientific antecedents, as well as by the development of prostheses to fill the holes created by the repression operations. Thus, as shown in Figure 4.1, Galileo's head has been transformed into a star and his phallus replaced by a cylinder with eyes. Significantly, in science, the eyes themselves are considered "male" organs, for in contemporary neuroanatomy, the eye is defined as part of the brain not the body, and it is interpreted as a central nervous system structure that penetrates to the surface of the body through holes in the skull called *eye sockets*. This choice of word is not fortuitous, for the double-headed organ, when turned on the stars, forms a circuit between heaven and earth, such that starlight passes down the tube, into the eye, and finally into the brain, where it is transformed—not into human experience—but into a new, objectified category of knowledge called *data*. Moreover, the myth tells us that these "male" body parts are unilateral, for only one eye is used when looking through a telescope and only one hand is used in writing and drawing—and we know which hand too, for in

English *write* is pronounced *right*. In confirmation of this mythic diagnosis, neuroanatomy tells us that the right hand is controlled by the left cerebral hemisphere, which is also the language hemisphere in the vast majority of humans, the so-called "dominant" hemisphere.[7] Thus, the symbolic system of science defines a pathway that runs from celestial bodies into the eye, through the left hemisphere of the brain, and back out to the right hand. The symbolism of the eye, however, is a little trickier, for neuroanatomy also tells us that in the human brain the optic nerve splits at the optic chiasma, such that the medial portion of the retina projects to the same side of the brain, whereas the lateral portion projects to the opposite side of the brain. For example, in the case of the right eye, the lateral portion, namely the part closest to the right side of the body, projects to the left hemisphere of the cerebrum whereas the medial portion, namely the part closest to the midline of the body, projects to the right hemisphere. Thus, in the body, as in the telescope, there are no left and right eyes, only higher and lower ones: the highest eye is the one that conveys starlight to the dominant hemisphere.

Thus, if we take a purely anthropological approach to Galileo's achievement, giving imagery and body language an equal place with theory and experiment, then the result is quite at variance with the official interpretation of the scientific method. As shown in Figure 4.1, an essential aspect of science is what I call the *Galilean circuit*, the channeling of celestial light through the eyes of a man-made phallus to create an enhanced image of nature accessible to a wider public. Although these enhanced images are now known as scientific discoveries, the primary function of science is not to describe natural events but rather to construct an image of them by means of lenses and celestial light. Natural events that cannot be transformed by photography and instrumentation are simply excluded from scientific consideration on the grounds that the phenomena are "too complex" or the terminology "too vague and imprecise." In fact, the image of the apparatus is far more important than the images of nature, which are only secondary and derivative. As the charter myth of science emphasizes, Galileo first used his telescope to look at ships at sea and only later applied it to astronomy, so it is the *apparatus* itself, not the facts of nature it reveals, that first amazes the merchants of Venice and makes Galileo into a star.

Thus, as shown in Figure 4.1, the Galilean circuit is the replacement of the scientist's own flesh and blood by a man-made phallic apparatus, such that data flow replaces blood flow. Because the Galilean circuit is in effect a kind of disembodiment, it is not surprising that contemporary

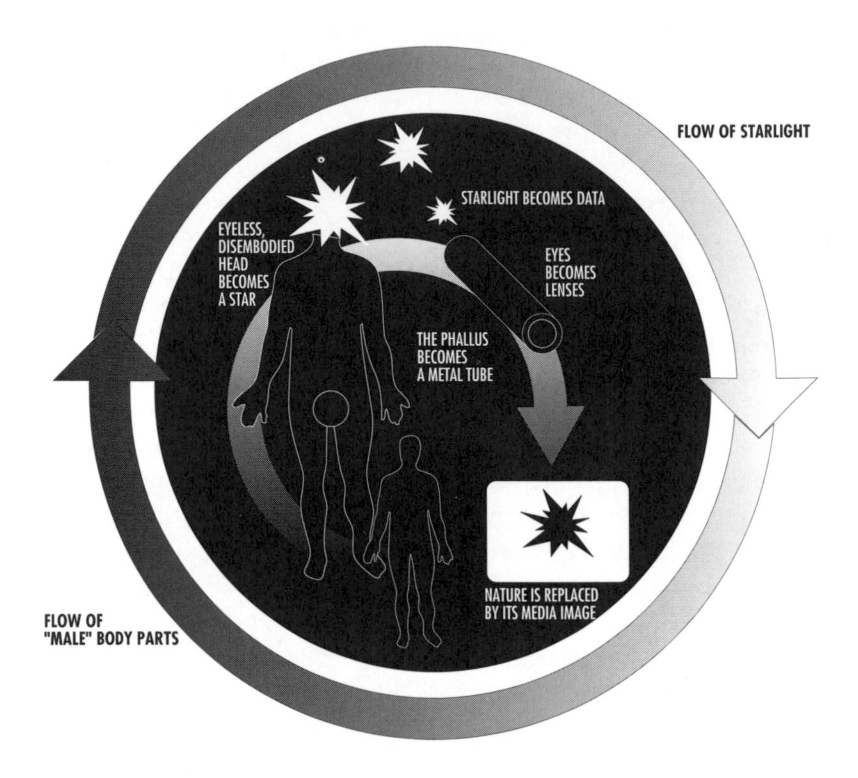

4.1 In the Galilean circuit, so-called male body parts are replaced by artificial organs, which channel starlight to earth.

astronomy takes this process to its logical conclusion and replaces the human eye with a photographic plate. On Mount Palomar, as Preston tells us, the light that passes down the telescope is normally used to expose a circular piece of film. That is, really scientific astronomers observe *pictures* of stars under artificial light, while presenting to the public at large huge, phallic apparatuses erected on mountain tops. If this process results in "discoveries," that is, in new, transformed images of natural events that capture the public imagination, then both the hero and his instrument are themselves subsequently transformed into celestial objects.

In the scientific world view, there is a profound separation between the "natural" and "symbolic" worlds, such that stars exist "out there" in nature whereas the names of stars are taken to be arbitrary cultural constructions, but empirically both the images of stars and their scientific names are symbolically mediated entities that are part of the same cultural complex; and the names of stars are as much a part of the scientific process as the stars themselves. This explains why scientists literally fight for the right to name a new image of nature. Significantly, in the modern, canonical sky chart, the person of Galileo is represented as the Galilean satellites of Jupiter, and his instrument has been transformed into the constellation Telescopium. The lenses of the instrument, Galileo's eyes, have been celestialized as well. The word *telescope,* Greek for "seeing at a distance," was first conferred on the new invention at a banquet held for Galileo in 1611 by one of the world's first scientific societies, the Accademia dei Lincei in Rome, "the society of lynxes." *Lynx,* which means "shining eyes," is also a constellation, recognized by its pair of bright stars, in the northern hemisphere between Auriga and Ursa Major. Not surprisingly, Lynx (7 to 9 right ascension, 30 to 60 celestial latitude, north) and Telescopium (19 right ascension, 45 to 60 celestial latitude, south) are at almost exactly opposite points on the celestial sphere, just as they are at opposite ends of the Galilean circuit.

In the story of Galileo and his telescope, which is set in Renaissance times, nature is viewed through a single eye from a fixed position, and this stance is also the defining characteristic of Renaissance art. As Robert D. Romanyshyn shows in *Technology as Symptom and Dream*, the invention of perspective drawing in 15th-century Italy makes it possible to convert a three-dimensional scene into a two-dimensional representation in a mathematically exact way. In one of the first descriptions of this technique by Leon Battista Alberti, published more than a hundred years before Galileo's telescope, the artist tells us to imagine the drawing surface as a window through which the scene to be painted is being viewed. The artist assumes that the viewer is standing on the horizontal plane and staring straight ahead through the "window," such that his line of sight converges on a "vanishing point" on a distant horizon line. Under these conditions, the horizon line in the painting is at the same level as the viewer's eye, and all the objects in the picture are arranged so that they will appear "in perspective" when viewed from this position. As art historians have pointed out, there is no picture known before 1425 A.D. that intentionally shows this construction of the relationship between viewer and subject matter. Romanyshyn rightly concludes that the Renaissance view of the world, as manifested in its art, is essential in the development of the technocratic body image: "Esconced behind the window the self becomes an observing *subject,* a *spectator,* as against a world which becomes a *spectacle,* an *object* of vision."[8] Moreover, he concludes, "the condition of the window also initiates an *eclipse of the body*,"[9] for the world is primarily something to be seen in two dimensions, not touched, smelled, or experienced. When nature is viewed from behind a window, it becomes equivalent to light. Thus, the technique of perspective art, which was developed simultaneously with the heliocentric theory, contains, as Romanyshyn points out, the three major ingredients of the modern technocratic world view: "There is the window as the boundary of separation between the viewer and the world; there is the self, portrayed in the figure of the artist, on this side of the window looking out on the world; and there is the world, out there, on the other side of the window, which has become primarily an object of vision."[10]

Moreover, in Alberti's original description of perspective drawing, the window is not open but literally draped with a veil, which divides the world into horizontal and vertical rows like a chessboard. As Alberti himself tells us: "A veil loosely woven of fine thread, dyed whatever color you please, divided up by thicker threads into as many parallel square sections as you like, and stretched on a frame. I set this up between the eye and the object to be represented, so that the visual pyramid [that is, rays converging on the vanishing point] passes through the loose weave of the veil."[11] Thus, as Romanyshyn points out, the veil that obscures the truths of nature is imposed by the Renaissance observer himself, who uses it to divide the unity of the original scene into arbitrary geometrical shapes, then strips it off in the name of science.

It is no accident, as Romanyshyn notes, that the same year, 1543, that the human body was ejected from the center of the universe by the theory of Copernicus it first appears as a corpse on the dissecting table of Vesalius[12]—for the very same acts of repression that transform the intellect into starlight also require that the abandoned body be incorporated into the system in objectified form. Thus, science, like industrial technology, is a one-two punch, in which the "head" of the body is projected onto the heavens, while the polluted, post-cranial remains fall back to earth, where they become grist for a whole new cycle of transformation by celestial light.

The disembodied observer viewing nature through a window is common to both science and Renaissance art, but, unlike the images of art, those produced by science are considered to be true descriptions of nature and not mere acts of imagination. Where Leonardo's painting *Lady with a Stoat* is firmly in the realm of "culture," Galileo's watercolor renderings of the phases of Venus are just as certainly descriptions of "nature." How can the same methodology lead to two such radically different interpretations? The answer is that science and art cannot be distinguished by their methodologies or by the images they produce but only by their *structural relationships to the human body on the one hand and to religion on the other.*

π in the Sky

The structural opposition between science and religion is encoded in the story of Galileo's censure by the Church for advocating the heliocentric theory of Copernicus, which is portrayed in popularizations as contrasting the unprovable tenets of theology with the empirical methods of science. For example, under the entry "Galileo" in *Webster's New World Dictionary*, we find that he improved the telescope, demonstrated the truth of the Copernican theory, and was condemned for heresy by the Inquisition. Or as Bronowski tells us: "By turning his telescope on the moon, on Jupiter, and on the sunspots, he [Galileo] put an end to the classical belief that the heavens are perfect and unchanging, and only the earth is subject to the laws of change."[13] James Burke in the television series *The Day the Universe Changed* is even less circumspect: "The telescope radically altered man's view of his position in the universe. No longer was he the centre of things, the uniquely chosen manifestation of God. The telescope freed man to look at himself and the world around him."[14] Moreover, Burke presents Copernicus as a modern scientist whose "principal aim was to explain the apparent anomalies in the motions of the planets with a simpler version of events than was currently held, closer in concept to the original plan adopted by Aristotle."[15] Thus, in canonical versions of the myth of The Lone Galileo, a physical act of observation proves the truth of the Copernican theory and repudiates a speculative theology; but, unfortunately, as modern scholarship has demonstrated, Copernicus was not a scientist, and the telescope was irrelevant to proving his theory.

Copernicus's book on the heliocentric theory is commonly taken as the founding date of science (1543 A.D.), but historical scholarship shows that its author, far from being the exemplar of the scientific method that he is commonly imagined to be, was a philosophical idealist whose vision of the world would be unrecognizable to a modern astronomer. Copernicus had been much influenced by Platonism, particularly by a collection of works in the Neoplatonic tradition that had recently been translated from Greek into Latin. The authors were Greek philosophers writing at the time of the Roman Empire, but they attributed their works to Hermes Trismegistus, who was equated with the Egyptian god Thoth. Unaware of this literary convention, Renaissance intellectuals thought Hermes was an ancient Egyptian philosopher from whom Plato derived the essential features of Platonic philosophy.[16] A hundred years after Copernicus's death, in the 17th century, linguistic scholarship would reveal that "Hermes's" works had been written in late Roman times and were completely derivative of Plato; but by then the damage, so to speak, had already been done. Copernicus thought he was in touch with the ancient roots of European philosophy. Reflecting on the Hermetic doctrine, he reasoned that as God is the source of light and life, then the sun, not the earth, is more logically the center of the universe. Applying this idea to astronomy, he showed that the motions of the heavenly bodies are consistent with the heliocentric conception. This heliocentric idea, far from originating with Copernicus, was, as he himself admits, well known to astronomers in the ancient world. Thus, Copernicus more properly marks a chapter in Platonic philosophy than in the history of modern science, and he is only regarded as a scientific ancestor because Galileo proclaimed himself a descendant. Moreover, this whole process had been set in motion by the naive acceptance of a literary convention as historical fact.

Indeed, the heliocentric theory, at that time, made no better predictions than the astronomy it sought to replace; and epicycles were still needed to explain certain movements of the heavenly bodies, just as with the system of Ptolemy. Moreover, there was no physical evidence for the motion of the earth at all.[17] Thus, in the 17th century, the Copernican theory was hardly a clear-cut case of fact versus superstition, and not even the astronomers could agree. Kepler was predisposed to the geometrical simplicity of the heliocentric system, but Tycho Brahe, whose meticulous sky charts helped confirm Kepler's laws of planetary motion, was an outspoken critic of Copernicus, and he offered a compromise of his own, in which the sun and moon went around the earth while the other planets went around the sun. Nor was the Catholic Church opposed to astronomy. In 1582, it reformed the calendar, using the services of the most eminent astronomers of the day, and it did not even oppose the teaching of the heliocentric system provided that it was taught as a mathematical device and not as a physical description of the heavens.

But in the popular mythology of science, the heliocentric theory of Copernicus is conflated with Galileo's use of the telescope, such that the invention of the latter proved the truth of the former. Although Galileo makes passing mention of Copernicus in *Sidereus Nuncius*, his translator

Van Helden points out that the radical aspect of Galileo's book is the role given to mathematics and instrumentation in the evaluation of astronomical theories. In Galileo's view, mathematics expresses a true description of physical reality, and instrumentation creates new sources of data that are simply unavailable to the naked eye.[18] Van Helden sums up this argument as follows: "Methodologically speaking, this was a very bold claim, for not only was there no optical theory that could demonstrate that the instrument [the telescope] did not deceive the senses, it was not even accepted in principle that optical theory could have much to do with reality."[19] In other words, Galileo was defining a new concept of truth, one based on an apparatus that he happened to control, that in effect repudiated the concept of truth, namely tradition and the scriptures, on which Christian dogma was based. It was not Galileo's astronomy but his philosophy that brought him into conflict with the church—and this constitutes the second phase of the myth.

The Lone Galileo, Verse Two:
Scientific discovery invalidates Church doctrine
And establishes the truth of a heretical theory;
But it leads to the trial of the hero
Who repudiates his heretical ideas,
Is banished from the public arena,
Loses his eyesight,
And writes his greatest work.

Galileo wrote his major defense of the Copernican theory in 1632, more than twenty years after his initial discoveries with the telescope; and shortly thereafter, the Church department charged with investigating heresy ordered Galileo to appear in a court of law, showed him the instruments of torture, and forced him to sign a document that repudiated the Copernican theory. The trial of Galileo is always faithfully recorded in popularizations of science, but the authors typically ignore the political aspects of the story, focusing instead on the fine points of difference between the mathematical models of Ptolemy, Copernicus, and Tycho Brahe. If they examine the institutional environment of Galileo's actions at all, they usually retrofit it with modern ideas of freedom of speech and technocratic notions of scientific autonomy; but as Redondi argues in *Galileo Heretic*, the Copernican theory, far from being the substance of the case, may have only been the legal vehicle that the Church used to silence Galileo. The proceedings were held in secret, and scanty documentation survives, so the exact nature of the

Church's grievances can never be known; but there is circumstantial evidence that the newly awakened concern in 1633 over Galileo's beliefs had more to do with the doctrine of the Eucharist than with astronomy, for the atomic theory advocated by Galileo, by repudiating the Aristotelian distinction between substance and accident as real principles of matter, vitiated the reconciliation of the doctrine of the Eucharist with natural reason.

The Mass is a ritual reenactment of the Last Supper of Jesus with his disciples, and essential to the ceremony is the sacrament of the Eucharist, in which the priest, by repeating the words that Christ used at his Last Supper, "This is my body...This is my blood," changes ordinary bread and wine into the actual body and blood of Christ. The mystical transformation of bread and wine is known in Christian theology as the doctrine of transubstantiation. In Renaissance times, the Church intellectualized this doctrine in terms of Aristotle's theory of matter, in which the bread and wine retained the appearance ("accidental" properties) of foodstuffs but were changed on the level of substance into the body and blood of Christ.[20] As shown in Figure 4.2, this physical transformation is paralleled by a spiritual transformation of the participants in the ritual. Just as the bread and wine are transformed into the body and blood of Christ while retaining their normal material attributes, so the bodies of the participants, while retaining material form, are themselves incorporated into a mystical body that embraces the entire Christian community.

Figure 4.2 also shows what happens when materialist reductionism is incorporated into this ritual context. By asserting as he did, that substances such as bread and wine have the physical attributes they do because of underlying differences in their atomic structure, Galileo in effect eliminates any possible disparity between the physical form of a substance and its observable properties. Thus, the atomic theory operates as the logical inversion of the Catholic Mass, reducing the body and blood of Christ to the material substances of bread and wine and precipitating the mystical body of the Christian community into a collection of individual physical bodies with no spiritual connections among them at all. Moreover, the agent of this transformation, the scientist, has the same role in the ritual of materialist reduction as the priest has in the ritual of the Mass, mediating between symbolic powers of transformation on the one hand and the community of the faithful on the other.

In short, the Church's instincts in regard to Galileo were sound, for by rejecting the Aristotelian distinction between substance and accident,

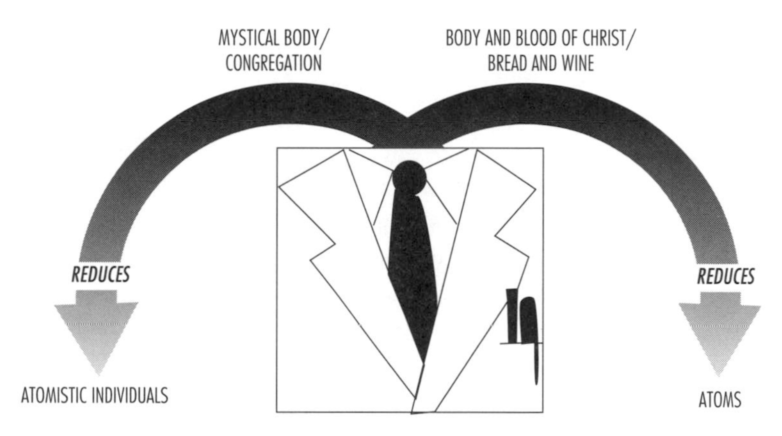

Figure 4.2 Scientific and Judeo-Christian rituals are in structural opposition.

the new physics threatened the very heart of the Catholic system of myth and ritual, the Mass; and it played into the hands of Protestant critics of the Church, who rejected both the doctrine of transubstantiation and the need for priests as mediating agents between God and the congregation. In Redondi's view, rather than fuel the fires of heresy by calling attention to these issues publicly, the Church tried to discredit Galileo by declaring him a heretic—not for the newly vulnerable doctrine of the Eucharist but on a legal technicality that linked him to Copernicus, the very Copernicus who was already condemned by no less a Protestant than Martin Luther himself.

As a condition of his punishment, Galileo publicly recanted his Copernican beliefs, and the Holy Office sentenced him to life imprisonment, which was subsequently commuted to house arrest by the pope. Although prohibited from advocating Copernicanism, Galileo was free to pursue more terrestrial research; and during the last years of his life, while under confinement, he wrote his greatest work, *Discourse on Two New Sciences*, which laid the foundation for modern physics. But paradoxically, in the surface structure of the myth, it is the Church that is the clear winner of the conflict, for it is Galileo who recants, is banished, and goes blind. He even had considerable trouble publishing *Discourse on Two New Sciences*, as no printer in a Catholic country would touch it. In short, Galileo dies, as he himself expresses it, "incurably blind, so that this heaven, this earth, this universe ... are now shrivelled up for me into such a narrow compass as is filled by my own bodily sensations."[21]

Thus, in the story, Galileo dies ignorant of the final verdict of history, which has to be provided by the commentator, in a meta-myth, who tells us that Galileo's last work is indeed immortal, dividing forever the intellectual worlds of Catholic and Protestant Europe, of theology and science. As the story of Galileo is remembered by people who are not otherwise interested in the history of science, who may even be unaware of Tycho Brahe's critique of Copernicanism, of Galileo's own crackpot theory of comets, or Cardinal Bellarmine's essentially 20th-century view of the role of mathematics in physical description, it must be because they remember the story of Galileo as myth and not as history. But if this be so, then why do they remember a myth in which it is the Church who emerges the winner? Although one can argue that the Church does not *really* win, because Galileo becomes a hero whereas the Church becomes increasingly irrelevant to the intellectual development of industrial society, the myth does not actually say this. This conclusion may be implicit in our modern reading of the historical events, but it is not itself a statement about the life of Galileo, who died, not only blind, but without four centuries of hindsight. Thus, in the classic version of the tale, Galileo dies under rather unfortunate circumstances, like a Hamlet or an Oedipus. His pathetic demise is only redeemed by the fact that he is ultimately vindicated by history, like a scientific Joan of Arc, who was burned at the stake as a heretic, only to have the judicial verdict reversed long after her death and the bones of her inquisitor exhumed and dishonored. Since history has in any case been edited, why does not science make the life of Galileo more consistent with its premises?

As the story of Galileo is always told in the context of Nebuchadnezzar's Dream, in which the Copernican theory is vindicated by history, his censure by the Church has the effect of making religion look ridiculous by implicitly contrasting outmoded superstitions to modern scientific truth. The myth is structured in such a way that a modern mind draws this conclusion even though there is nothing in the overt description of events that explicitly makes reference to it. Although the trial subsequently proved to be a strategic political blunder, for it provided historical material for a persuasive anticlerical mythology, modern scholarship has shown that there is nothing inherently ridiculous in the Church's views on the Copernican theory. The Church did not deny the truth of Galileo's observations; to the contrary, the Jesuit astronomers at the Collegio Romano were among the first to verify his findings. The Church was not hostile to astronomy; it had hired astronomers to reform the calendar. The Church did not believe in the literal interpretation of the Bible; that is a Protestant contribution. The Church was not resistant to change; it was engaged in an extensive program of reform. Moreover, in the years before the trial, the Church *did* allow discussion of the Copernican model among astronomers, provided that it was taught as a mathematical construction and not as a physical fact. (Nor is the Copernican theory taught as fact today: few 20th-century astronomers would claim that the sun is the center of the universe and immobile.) Moreover, the strongest evidence for the heliocentric theory was its mathematical simplicity, which owed more to Kepler's laws of planetary motion and Newton's laws of gravitation than to Galileo's observations through the telescope. However, when people in a technocratic society are presented with the skillfully edited facts of the Galilean myth, they draw the conclusion that the Church is ridiculous when compared with science.

The second function of the myth is to show that Galileo is immortal in spite of his ignominious death. This also is accomplished by embedding the story of Galileo's trial within Nebuchadnezzar's Dream, so that his improvement of the telescope marks the beginning of a long period of astronomical progress culminating in the landing on the moon. This interpretation of history not only firmly establishes religion as a more primitive stage of human thought, but it also allows the spirit of Galileo to be carried forward after his death in a secular rendering of immortality.

The two verses of The Lone Galileo, namely Verse One, in which Galileo makes a number of discoveries about the nature of the universe, and Verse Two, in which we learn of his personal fate, roughly correspond to the difference between what theologians call the cosmogonic and eschatological functions of religion. *Cosmogony* answers the question "How did we get here?" whereas *eschatology* answers the question

"Where are we going?" The cosmogonical and eschatological functions, far from being incidental to science, are the essence of the system, for it is these that motivate scientific practitioners to devote their lives to the enterprise and establish scientists as the spiritual leaders of our time. In fact, the cosmogonic and eschatological functions dominate the endless popularizations of science that are broadcast on American television. In Bronowski's *The Ascent of Man*, for example, the immortality of scientists is unequivocally expressed: "Physics in the twentieth century is an immortal work. The human imagination working communally has produced no monuments to equal it, not the pyramids, not the *Iliad*, not the ballads, not the cathedrals."[22]

The immortality of scientists is not just a vulgar affectation of mass-media popularizations but, as the charter myth itself suggests, is an essential aspect of mainstream science as well. In the dedication to *Sidereus Nuncius* Galileo names the four moons of Jupiter the "Medicean stars" after his patron Cosimo II de' Medici; and he explicitly points out that whereas all forms of earthly immortality inevitably wane, he can offer his patron a fame that will endure until "the splendor of the stars themselves is extinguished."[23] He assures the prince: "for scarcely have the immortal graces of your soul begun to shine forth on earth than bright stars offer themselves in the heavens, which like tongues, will speak of and celebrate your most excellent virtues for all time."[24] Operationally defining an immortal as someone whose framed portrait is hung in an institution after he is dead, then science is still as addicted to stars as Hollywood. Indeed, the walls of the theoretical astrophysics division at Caltech are decorated with photographs of Niels Bohr, J. Robert Oppenheimer, and Albert Einstein;[25] and in scientific textbooks, there are always pictures of famous scientists interspersed with the equations to show the student the clear and unambiguous relationship between the mastery of scientific content on one level and the proof of star quality on the other. Although in the folk theory of science, stars are considered one thing, their names another, to the anthropologist the celestialization of the scientist's body is an essential component of the Galilean circuit, for it connects the dry, disembodied products of scientific research with such normal human motivations as the desire to transcend death. Indeed, to achieve immortal status, the scientist need only emulate the life and works of the culture hero himself—for are not the Medicean stars now known as the Galilean satellites?[26]

The Two Faces of Science

Because eschatology is as important to science as it is to theology, the two institutions are necessarily in competition for the same niche, and this explains why the trial of Galileo gets so much air time in contemporary America, a culture not otherwise known for its historical depth. The Lone Galileo is 20th-century myth not 17th-century history, and tells us that religion is a *prerequisite* to *contemporary* science, whose charter is to demonstrate physical facts (Verse One), then use them to repudiate a theological position (Verse Two). Therefore, to faithfully follow in the path of Galileo is to challenge Church doctrine by demonstrations of physical facts that seem to contradict theological propositions. Far from trying to abolish religion, science seeks to preempt it, assuming the mantle of theology for itself.

Some religions are more congenial to this scientific agenda than others. In mainstream Protestant sects, the ones that founded the Ivy League universities, science and theology are considered to be oars on the same boat, with science responsible for matter, theology for the spirit. Progressive theologians, taking science at its word, as an open-minded inquiry into the laws of nature, and fancying themselves as the second oar, like to see scientists as closet Christians.[27] Much is made, for example, of the lecture entitled *Religion und Naturwissenschaft* ("Religion and Natural Science") delivered in 1937 by the physicist and "dedicated churchman" Max Planck,[28] which affirms the importance of theology in the modern world by telling us that Kepler, Newton, and Leibnitz were very interested in religion. Even the vague, pantheistic rambling of Einstein is somehow construed as endorsing the Judeo-Christian concept of God. But nature theology fails to see that this open-mindedness is a one-way street, for scientists have no reservations whatsoever in appropriating theology's turf whenever they have the power to do so. Also, it confounds the beliefs of individual scientists with the role of the institution. Even if hitherto undiscovered documents prove conclusively that Einstein was a practicing Presbyterian, it will simply underscore the fact that personal religious practices are not definitive of the scientific agenda—Galileo went to church too. Not only is there a profound intellectual objection to the whole idea that science and theology have a natural division of labor between body and soul, matter and spirit, but this doctrine ignores the covert agenda of the scientific program and the unconscious elements of its iconic deep structure. The analysis of science's own charter mythology indicates that the two disciplines are in structural opposition and essentially unreconcilable within the context of industrial society.[29]

Once again, there is an anomaly in the myth of Galileo, an anomaly rarely pointed out, and never by the popularizers: the fact that the founding hero of science is not very heroic. Where John Wayne would have faced down his accusers with manful, monosyllabic contempt, and Rambo blown off the doors of the papal palace, Galileo signs the document retracting his vision of the truth. Even from the vantage point of history, Galileo's predecessors were made of sterner stuff: Bruno preferred to burn at the stake rather than repudiate his Neoplatonic theology, and Sir Thomas More tipped the executioner with a gold coin before resting his head on the block. But Galileo probably avoided a similar fate by signing a document that said he repented of his heresy and repudiated Copernicanism, agreeing that he "must altogether abandon the false opinion that the sun is the centre of the world and immovable, and that the earth is not the centre of the world, and moves."[30]

Did Galileo *really* repudiate his Copernican ideas? The implication of the popularizations is that he did not, for otherwise he would have suffered in vain, without even the cold solace of scientific immortality. Yet consider the implications of his actions, if not for Galileo, then for his intellectual descendants. If the founding hero of the scientific enterprise can sign a document that in effect repudiates all that he believes in, what does this tell us about the scientific concept of truth and the relationship of science to authority? It says that scientific heroes should lie about their beliefs if their activities ever bring them into conflict with governments and theologians, even if it means presenting a public posture that is in conflict with scientific truth. This mythology also shows why scientists are so useful to the state, for even when they are privately skeptical about religion and politics, they can be publicly counted on to do whatever is necessary when the chips are down, such as Oppenheimer, who literally buried his pinko past to heed his country's call. The charter myth of science says to its practitioners: "Don't be heroes"—and not surprisingly, not one of them is.[31]

There are two major public events in the life of The Lone Galileo, namely his demonstration of the telescope and his public capitulation to the inquisitors, and this suggests that the institution of science has a double standard of truth: a firm commitment to truthfully expressing empirical knowledge supplemented by a willingness to deceive organized religion as to the nature of its real agenda and beliefs. Moreover, Verse Two of the charter myth begins with a challenge to the Church, suggesting that this course of action is equal in value to the practice of science in a strict sense of the term. Also, the Copernican theory, far from being neutral, had already been censured by the Church, for it was suspected, probably justly so, of being a heretical system of theology disguised as astronomical propositions. In other words, the tale of the Lone Galileo is not only a blueprint for a complete religion, with both cosmogonic (how we got here) and eschatological (where we are going) functions, but it encodes a practical program for challenging theological propositions while denying that this is really one's intention at all.

In scientific biographies, this phenomenon is typically expressed as a conversion to science after a crisis of faith. For example, in *The Making of the Atomic Bomb*, Rhodes tells how the young Einstein while attending school took a foray into Judaism, only to conclude that many stories in the Bible could not be true. As Einstein expressed it, "'Suspicion against every kind of authority grew out of this experience, a skeptical attitude towards the convictions which were alive in any specific social environment.'"[32] This quotation is interesting, not only because Rhodes takes it at face value, but because Einstein is held up to us, more than anyone else, as the modern scientific ideal: the pacifist genius of vaguely leftist leanings dedicated to enlarging the intellectual horizons of all mankind. But the problem is that Einstein's pacifism and self-proclaimed skepticism towards all forms of authority did not extend to burying Szilard's proposal for an atomic bomb under the pile of papers on his desk. Instead, he used his own considerable moral and intellectual authority to launch the largest weapons development project in all of recorded history—the Galilean hero par excellence.[33]

The theological referents of scientific activity are never mentioned in the "technical" scientific literature, only orally and in the popularization; and if pressed on any of the underlying religious and philosophical issues, scientists, like the censured Galileo, disclaim any theological intent. Nonetheless, almost all popular works on science, as well as memoirs by scientists themselves, convey an implicit theological agenda in which manifestly metaphysical propositions, such as free will or the perfectibility of human nature, are evaluated in the light of scientific theories of nature—and found to be wanting. For example, a 1989 cover story in *California* magazine on the Caltech astrophysicist Kip Thorne describes how he mathematically models billiard balls traveling backwards in time by means of "wormholes" in the fabric of space-time.[34] On the surface, this is pure science, because it is overtly concerned with billiard balls, relativity, and physical laws, but further along in the article we come to an atheological conclusion: Thorne asserts that his research into billiard balls reveals contradictions in the concept of free will.[35] What sort of mind investigates billiard balls in order to understand free will? The pervasiveness of science in a culture is measured by the extent to which theological and philosophical questions are confounded with the mathematics of gravity—and nobody notices.

The bifurcation of scientific thought into a seemingly neutral theory expressed in textbooks and a vociferous theological position expressed in so-called popularizations has been part of the scientific enterprise for centuries. The discovery of atomic energy, for example, exemplifies this strategy perfectly. In 1901 Ernest Rutherford and Frederick Soddy discovered that an atom of thorium changes into an atom of argon by the emission of radioactivity, and they carefully examined other radioactive elements in turn, establishing their breakdown products and measuring their half-life. Soddy's scientific description of the transmutation of elements, for which he won the Nobel Prize, has all the dispassionate precision of pure science, but in his popular works he attacked the Judeo-Christian doctrine of original sin. In his widely read *The Interpretation of Radium*, published in 1908, he wrote that "A race which could transmute matter would have little need to earn its bread by the sweat of its brow.... Such a race could transform a desert continent, thaw the frozen poles, and make the whole world one smiling Garden of Eden."[36] Moreover, Spencer Weart, himself a physicist, in *Nuclear Fear: A History of Images*, shows that the articulation of all of the major policy issues pertaining to the use of atomic energy—underground shelters, limitless energy, universal holocaust, and universal peace—*preceded* both the scientific discoveries of atomic physics and the development of atomic technology, indicating that the mythical concerns, far from being post hoc rationalizations, are in fact the driving force behind the scientific and technological activities. In the same vein, historians of science have shown that Isaac Newton was a closet alchemist who believed himself to be the divinely chosen representative of the ancient Hermetic wisdom, and he even rewrote the Latinized form of his name, Isaacus Neuutonus, as the anagram *Jeova sanctus unus*—Jehovah the holy one.[37] But none of the alchemical and occult sources of his thought are referenced in his published works, which present his conclusions in terms of a sanitized mechanical philosophy.

Yet the academically orthodox position is that of Stephen Toulmin, a respected philosopher of science, who draws a hard-and-fast distinction between the truth of scientific theory and the distortions imposed by the myth-making process. He tells us that the 20th century considers itself myth-free because "the material from which we construct our myths is taken from the sciences themselves,"[38] so that modern myths are mechanomorphic, not anthropomorphic. He also points out that this fact shows the impossibility of the Enlightenment's dream of a scientifically grounded system of economy and government, for to the extent to which a scientific term is used metaphorically to justify the wider concerns of society, it loses the relationship to a well-defined class of physical phenomena on which its scientific precision is based. Yet Toulmin's argument still assumes that scientific activity is itself essentially myth-free, a generator of truths, so that *myth* is something that happens to scientific knowledge in the course of its journey from the ivory tower to the factory floor.[39] Philosophers and historians of science are themselves professional academics, with a vested interest in the concept of the expert who commands the machinery of truth, so they take at face value the major premises of science itself: first, that its first priority is truth; second, that it produces objective descriptions of nature; third, that physical theory is the engine of the enterprise; and fourth, that mythic conceptions have been rigorously excluded from the fundamental operations of the scientific process. In effect, it is science itself that defines the proper subject matter of academic inquiries into science, declaring the attic and the basement off bounds to inquisitive sleuths.

The Atheological Agenda

Yet the interpretation of science as a covert, atheological agenda can explain major conceptual changes in the history of Western thought without recourse to science's own ideology of theoretical progress. Ever since the work of Thomas Kuhn on *The Structure of Scientific Revolutions*, scholars have been aware that science advances through what this scholar calls paradigm shifts—the replacement of one scientific world view by another. The new paradigm provides an alternative interpretation of previously established scientific facts, introduces new concepts and research techniques, and, most importantly, defines the problems that still need to be solved. The paradigmatic paradigm is Einstein's theory of relativity which replaced the three-dimensional space of classical mechanics, itself conceptualized as a kind of empty box in which material bodies interacted through forces at a distance, with a radically new vision of physical reality in which matter bends the fabric of four-dimensional space–time to create gravitational effects undreamt of by Newton. But as Kuhn points out, before a paradigm can become accepted, "the new candidate must seem to resolve some outstanding and generally recognized problem that can be met in no other way. Second, the new paradigm must promise to preserve a relatively large part of the concrete problem-solving ability that has accrued to science through its predecessors."[40] But significantly, as Kuhn expresses it, "paradigm debates are not really about relative problem-solving ability, though for good reasons they are usually couched in those terms. Instead, the issue is which paradigm should in the future guide research on problems many of which neither competitor can yet claim to resolve completely."[41]

Although there is no denying that new scientific paradigms typically address and resolve many of the problems raised by prior scientific research, Kuhn's analysis implies that the source of scientific revolutions is to be found within the intellectual framework of science itself. Thus, his interpretation, for all of its surface iconoclasm, reaffirms science's own folk model that the theory is where the action is. Moreover, it is not Kuhn's theory which determines what is and what is not a scientific revolution—it is science itself that dictates these judgements. So Kuhn's re-analysis of scientific development may jettison the dogma of mindless empiricism, but it also preserves the arrow of progress from Newton to Einstein, which explains, no doubt, why Kuhn is a tenured professor at Einstein's very own institute, the Center for Advanced Studies at Princeton. In short, Kuhn's theory of scientific change not only assumes science's own ideology of forward motion, but it elevates these folk interpretations to a theory of history in a strict sense of the term.

The advent of Einstein's theory of relativity is more satisfactorily explained by the hypothesis of a covert atheological agenda of institutional science than by anomalies in classical mechanics. Admittedly, there were intellectual problems with the previous paradigm, as there are in any system of thought, but problems do not create their own answers. Rather, the new paradigm is first evaluated for its atheological potential, and only after it passes this test is it systematically applied to scientific problems. In the atheological interpretation, Einstein's intense antipathy to organized religion, documented by his own statements quoted above, is the real inspiration of his physical theory, not the Michelson-Morley experiment. Moreover, it was Newton's theology, not his mathematics, that was becoming increasingly embarrassing to physics, for Newton had devoted a good deal of his professional life to mathematical explications of the book of Revelation; and his celestial mechanics presumed the actions of a Creator who imparted to the planets their initial motions. Although in Newton's cosmic drama God had been demoted from producer to set director, arranging the scenery against which subsequent physical actions take place, in Einstein's relativistic universe the planets are locked into their relative positions by purely physical forces, a curvature of space–time appropriate to their masses. Thus, the divine presence can be banished to the first few nanoseconds of the universe, making it possible for science to provide a complete and seamless cosmogony stretching from the act of creation to the lunar landing. It is the hegemonic potential of Einstein's ideas that made them so instantly attractive to 20th-century physicists.

Every generation of professional scholarship digs into the scientific mother lode to extract its own shiny nugget to characterize the uniqueness of science—empirical, experimental, progressive, falsifiable, to name only the most famous finds; and these examinations of the scientific process by academic outsiders have shown the inadequacy of the scientists' own explanations of what it is they do. Kuhn's historical research shows, for example, that science does not progress by the patient accumulation of facts and the judicious weighing of negative evidence. Instead, theories organize the perception of facts as well as their acquisition, so that observations that do not fit the theory are largely ignored until a new theory comes along that can better accommodate them. Moreover, individual scientists typically do not abandon the theory of their youth in favor of a radically different way of looking at the world, changing from Newtonians to Relativists, or from Catastrophists to Darwinians. As in other institutions, the old guard eventually dies off, abandoning the field to the young. Thus, historical and philosophical studies of the scientific method show that science, whatever it is, is certainly *not* a dispassionate observer staring wide-eyed at nature, building instruments, and writing down facts. Also, a "scientific fact" is not something that one can establish simply through direct observation, for the facts, like theories, are a collective effort. When Galileo looked at the planet Saturn, he did not see the concentric rings we now take as self-evident but described it as a sphere with a bump on each side, just as when Herschel first saw the planet Uranus with an improved telescope of his own construction, he interpreted it as a comet.[42] Yet, in spite of their contribution to our understanding of the scientific process, the historical and philosophical disciplines have also reinforced the scientific world view by assuming, with science itself, that theory is where the action is.

In Kuhn's analysis of the history of science, for example, when one paradigm stumbles, another paradigm picks up the pieces and carries on; but the recruitment of scientists to the new theory cannot itself be explained by the rules of scientific evidence. The validation of a new theory requires that a number of scientific researchers abandon the established framework to make observations and undertake experiments in conformity with the new guidelines, a process that requires commitment of time and resources in advance of the evidence needed to justify such a change. For this reason, the philosopher Paul Feyerabend, in *Farewell to Reason*, has likened the paradigm shift to a conversion process within science itself.

The classic case is that of Charles Darwin, whose theory of evolution by natural selection, as presented in 1859, did not resolve any pressing scientific issues, and had almost no scientific evidence to support it.[43] In fact, it was contradicted by a wide range of well-established scientific facts across a broad range of disciplines. By Darwin's day, vertebrate paleontology had developed into a significant science that had excavated the fossils of hundreds of extinct animal species, from mammoths to dinosaurs, and it showed that the history of life on earth is punctuated by mass extinctions in which one type of fauna dies off, to be replaced sometime later by another, a process that Cuvier called catastrophism. But Darwin's theory assumes gradual, incremental changes in species, analogous to the selection of characteristics by animal breeders, and it was flatly contradicted by the fossil record so laboriously built up by paleontology. Darwin argued that the gaps in the fossil record would eventually be filled in, so that species could be seen to merge one into another over time, but this optimism has not been borne out. (In fact, the essential features of Cuvier's catastrophic evolution, minus the doctrine of special creation, have been scientifically rehabilitated in the last decade under the new name of punctuated equilibrium.) Also, Darwin's theory assumes that the natural state of a species is genetic diversity, so that some individuals can be better adapted to their environment than others. Indeed, to Darwin, the very concept of a "species" is intrinsically statistical. As a breeder of show pigeons, Darwin had a good practical understanding of genetic selection, but there was nothing in the biology of his day to support the idea that species are intrinsically variable in nature. To the contrary, academic biology thought of species as ideal forms, like Platonic archetypes, a concept in which statistical variability is assumed to be irrelevant or even pathological. Moreover, a science of genetic diversity supportive of Darwin's conclusions would not become generally available to the scientific community until several decades after the theory of natural selection had become commonly accepted among educated people. Also, there were significant arguments against Darwin's theory made by more established sciences. As the gradual accumulation of adaptive characteristics required an extremely long time scale in which to work, the eminent physicist Lord Kelvin argued on the basis of physics that there was simply not enough energy in the sun to sustain the earth for the time scale required by the theory of natural selection. Thus, Darwin's theory, when first presented, lacked any evidence for innate variability; it was contradicted by the fossil evidence of the history of life; it assumed a statistical concept of the species quite foreign to the working biology of the day; it did not resolve any specifically scientific issues raised by previous biological research; and it appeared to be in conflict with physics. The best that one could say was that it was consistent with Lyell's uniformitarian geology, that it explained the similarity of embryos among widely divergent species, and that it resolved anomalies in the geographic distribution of plants and animals. Although these topics no more animated the average man of Darwin's day than they do now, the book was an instant best-seller; and the theory of natural selection became canonical science by the end of the 19th century—well in advance of the observations and mechanisms that are now invoked as evidence for it.

Significantly, Darwin's theory cannot even be called a paradigm by Kuhn's definition because it did not resolve the contradictions of previous biological theories, and it ignored the fossil record entirely. But it did effectively demolish the argument that Darwin *claimed* he set out to refute, namely, Archdeacon Paley's idea that the existence of God can be proved by the evidence of design in nature. Paley, an Anglican clergyman, had written a number of books on biological subjects that he used as evidence for an intelligent creator. His thesis is often expressed by the story of the lost watch: if you found a watch on the beach, the intricate design of the artifact with its multiplicity of parts working together would enable you to infer the existence of a watchmaker, even if there was no other evidence that watchmakers existed. To the theologian Paley, this was a forceful natural argument for the existence of God (what sort of God is another matter), and it is this thesis which Darwin debunked in his scientific writings, showing that the complexity of organisms developed gradually through the selective retention of adaptive characteristics, what he called the principle of natural selection, and that the design in organic nature so obvious to Paley was in fact largely illusory. In Darwin's case, where significant scientific evidence for his position was all but non-existent, this not-so-covert atheological agenda was the primary criterion for scientific acceptance of the theory, and it explains the enthusiastic reception by all factions of the industrialist intelligentsia: by Marx and Engels who saw it as proving their materialist theory of human nature, by Fabian socialists who saw it as the agency for the perfectibility of man, and by laissez-faire capitalists who thought it showed that unrestricted competition leads inevitably to progress. But Darwin's greatest service, still too little appreciated, is to theology itself—for he was as good as his word and banished the banal abstraction of Paley's divine watchmaker from God's green earth forever.

As with the presentation of established scientific theories to a general audience, the elimination of prescientific ideas that conflict with the underlying atheological agenda is done in the pages of the popularization. In fact, a prescientific idea judged to be atheologically anathema is personalized in popular scientific writing as a character found nowhere else but in this genre—the character of "the crank." In this literature, there are antievolutionist cranks, antivivisectionist cranks, antinuclear cranks, indeed as many classes of crank as there are critics of science's institutional position as arbiter of truth and heir to theology. It is for this reason that the most severe criticism is reserved for people who claim to be scientists but who refuse to accept the underlying atheological program, such as the universally condemned arch-crank Immanuel Velikovsky. The problem with Velikovsky is not with the intrinsic craziness of his ideas, for there are numerous notions in contemporary, mainstream science of such crystalline lunacy as to make Velikovsky's baroque excrescences seem drab in comparison: pollution-free sources of energy, providential asteroids exterminating dinosaurs, time-travelling through wormholes—to list only ideas on the cutting edge. Rather, the problem of Velikovsky is that he uses scientific data and methods to try to prove the essential truth of mythical conceptions of reality as preserved in religion and folklore—a research program totally at cross-purposes with everything that science stands for. No wonder he is called a *crank*—a device that converts one form of motion into another.

Thus, it follows that popularizers are the gatekeepers of the scientific enterprise, for it is in the popularization, not in the technical treatise, that mythical concepts are introduced, evaluated, and adopted. Although in academic circles *popularization* is a dirty word, suggesting the debasement of pure ideas by the values of the marketplace, the more important the scientist (as measured on purely scientific criteria, such as the number of journal citations and the dollar volume of research grants), the more likely the scientist is to write popularizations in a strict sense of the term, namely books and articles addressed to the general public: Watson's *The Double Helix*, Pauling's *Vitamin C and the Common Cold*, and so on.[44] This too has a mythical charter, for Galileo's career begins, not with a new and better theory, as academic historians and philosophers of science would lead one to expect, but with an unequivocal *popularization,* namely, the demonstration of the power of the telescope to the rich merchants of Venice. Until catapulted to fame by his exploits on the Campanile, Galileo was a respected university professor, popular with the students, but he had far less claim to immortality than many of his contemporaries. Only after his successful foray into the marketplace does the mythic Galileo begin to do scientific work in a strict sense of the term, such as observing heavenly bodies, publishing the results, and raising doubts about a predecessor's theory. In short, Galileo becomes a famous scientist *before* he does any memorable science.

The dramatic conflict between science and religion in the tale of The Lone Galileo indicates that science to this day has a theological agenda—*theological* in a strict sense of the term, namely, the study of God, particularly the relation between God, mankind, and the universe. The primary agenda of science is to promulgate its own materialist inversions of theological propositions (atheologies) at the expense of overt religion—and this is not surprising, given that Nebuchadnezzar's Dream is itself an inversion of a biblical myth. In other words, science is a negative to theology's positive, and by compiling a simple list, it is easy to show that each scientific theory inverts a thesis in Judeo-Christian theology while carefully skirting any real consideration of the underlying philosophical and existential issues:

Thesis	Antithesis
The world was created by God.	The world was created by the Big Bang.
Human beings have a spiritual as well as a material nature.	The world is made out of matter.
God is immanent in His creation.	The world is a machine.
God is a person.	Scientific laws are statistical.
People are responsible for their actions.	Human behavior is determined by genes and environment.
God loves his creatures.	Altruism is just genetic self-interest.

The latent agenda of the scientific enterprise is to *replace theological propositions by materialist propositions while denying that this is the intention at all.*

Thus, it is hardly surprising that each philosopher of science comes away with a different interpretation of the essence of the scientific method, for the distinctiveness of science is not in its method at all but in the underlying imagery of the transformation of nature by starlight and in the ideology of objectivity with which it disguises its atheological agenda. For all of the talk about evidence and empiricism, the facts and theories of science are like the pieces on a chessboard: place holders in a higher strategy. Although chess tokens are demonstrably *physical* entities, each with a certain weight, shape, and size that can be objectively described and measured, their material properties are irrelevant to the game, which can be played equally well on a computer screen or with a different set of conventional tokens entirely. In the same way, we accept the intellectual revolutions of our time—the shift from Newtonian to relativistic mechanics, from uniformitarianism to plate tectonics—with such equanimity because we already know in our heart of hearts that scientific content is irrelevant to our faith in the technocratic ideology. The geological record, the diversity of species, the topography of the sea floor, the development of galaxies, the neurons of the human brain—all of these natural phenomena are grist for science's mill, but none of them will have any more effect on our conception of the world than the shape of the tokens has on the outcome of a chess game. Just as in Nebuchadnezzar's Dream, in which constantly evolving tools and techniques affirm the eternal truth of the underlying mythology, so the frantic forward motion of scientific research disguises profound stability in the underlying cultural assumptions. Science does not so much "progress" as recycle its charter myths in new, more dramatic forms, so we never have to address the fact that the underlying mythology—Nebuchadnezzar's Dream and The Lone Galileo—has scarcely changed in four hundred years.

THE TWINS WHO BRING THE FIRE

The One-Two Punch of Physics and Biology

Physics is part of the superstructure.

Niklas S. Damiris[1]

Technology, far from being a neutral and mechanistic process, is embedded in a larger mythical framework, and its tools and techniques are ritual replications of the symbolic body. As shown in previous chapters, the symbolic body of postwar industrialism is defined by the myths of Nebuchadnezzar's Dream, Bloody Stumps/Exploding Stars, and The Lone Galileo. This world view is owned by the technocratic players and implemented by mechanists, who together create rituals of reduction that wrench the members of society from their traditional social and ecological contexts and fill the holes with artificial replacements built by technocratic methods—a process called the one-two punch. This world of reconstructed images is projected onto the landscape as "objective" descriptions of natural and social events that reinforce the premises that created the methodology in the first place. Thus, in the anthropological perspective, technology is defined as either a landscape projection of the symbolic body or, more narrowly, as the artifacts and physical techniques that implement such projections.

This unabashedly symbolic definition of technology is not only far more consistent with the archaeological evidence of human evolution than is Nebuchadnezzar's Dream, but it reveals aspects of modern technology that are completely obscured by conventional approaches. For example, the atomic bomb, that most representative of modern inventions, is portrayed in popular histories as a logical development from physical theory, and its construction is explained as a consequence of the military pressures of the Second World War. But closer scrutiny shows that the atomic bomb is primarily a ritual artifact that stands in a structural relationship to the biological body on the one hand and to the imagery of bloody stumps/exploding stars on the other. Once these correspondences are pointed out, the mythic imagery of the bomb is easily recognized by nonscientists, but it is difficult for technocrats themselves to admit because it cannot be verbally acknowledged without threatening the assumptions of the mechanistic world view.

In the technocracy's own folk history, science develops logically from demonstrable anomalies in scientific theory itself, augmented by daring acts of empirical discovery though the construction of new apparatus—a concept of knowledge that pointedly excludes any mythical constructs, covert imagery, or atheological agendas. Moreover, this cozy, self-sustaining world of scientific theory is bolstered by academic divisions of knowledge that rigorously separate the content of one field from that of another while ignoring the technical and institutional relationships that connect them. In histories of the atomic bomb, for example, professional physicists are invariably the protagonists, and scientific events that take place outside of physics are simply ignored.[2] Biology, for example, never appears in histories of the atomic bomb, for it is conventionally regarded as a different branch of science, remote from the story of nuclear weapons. Yet physics and biology not only have much in common when viewed in terms of the symbolic body, but in the 20th century they developed in concert, often using similar imagery and technology, and they were funded by the same sources. Moreover, the development of the atomic bomb was paralleled by an equally heroic effort in biology, namely, the mass production of penicillin; and both projects followed a similar time course, beginning in earnest about March 1943 and producing useful results by the middle of 1945. The rationale was similar too, for both the atomic bomb and the development of penicillin were justified in terms of their military applications, and both were joint ventures of the British and American governments.[3] But in histories of science, the Manhattan Project is dubbed "physics" and penicillin "biology"; the former is judged "military," the latter "humanitarian."[4] To the anthropologist, however, these two developments of penicillin and the atomic bomb are structurally related within a common symbolic system of thought; and far from being "separate scientific disciplines," each with a linear history of progressive development, physics and biology are in fact complementary processes of mutilation and prosthesis: the two fists of a one-two punch.[5]

Biology and physics are connected intellectually as well, on the level of physical concepts and experimental technique. Twentieth-century biology is largely the history of genetics and molecular biology, and the development of these fields was critically dependent on nuclear physics through the use of X-ray diffraction to explore the structure of large molecules, the use of radioactive isotopes to trace biochemical pathways, and the use of radiation to induce mutations. The sources of these materials and equipment were nuclear physicists themselves, and many of the protagonists in the story of the atomic bomb regularly crossed disciplinary boundaries. E. O. Lawrence was a strong supporter of radiation therapy for cancer, even irradiating his mother with his cyclotron at Berkeley, and he set up an office to distribute radioactive isotopes to medical laboratories all over the world. Leo Szilard, the instigator of Einstein's famous letter to Roosevelt advocating the development of the A-bomb, moved directly, in 1947, from working on nuclear weapons at Los Alamos to researching the structure of viruses. Linus Pauling, Watson's rival in the race to *The Double Helix*, first achieved scientific eminence by studying the nature of chemical bonds. And Erwin Schrödinger, a physicist, author of the eponymous wave equation that is one of the foundations of quantum mechanics, fired the starting gun in the 1930's with a series of highly influential lectures, published as the slim volume *What is Life?*

Significantly, the organizations that funded scientific research in the two decades that preceded the atomic bomb, in the late 1920's through the early '40's, made no hard-and-fast distinctions between physics and biology but saw them as mutually complementary endeavors within a larger ideology of material progress, rational control, and the conquest of nature. One of the major contributors to science at the time was the Rockefeller Foundation, which contributed many millions of dollars a year to basic research in both the United States and Europe.[6] The Rockefeller Foundation's support for basic research had developed gradually from a hodgepodge of Rockefeller family charities devoted to health, education, Baptist missions, and war relief, but in 1928–29, these organizations were consolidated and restructured to facilitate a new, more secular charter: the well-being of the human species through the advancement of scientific knowledge.[7] As one historian expresses it: "The reorganization of 1928 represented the final triumph of the scientific patron over the missionary in the battle for control of the Rockefeller philanthropies."[8] This new mission was consolidated in a new division of natural sciences, and in 1928 the foundation hired, for the first time in its history, professional scientists to disburse its scientific grants.[9]

The foundation made grants to well-established scientists and institutions, such as Oxford University and the California Institute of

Technology, and it sponsored traveling fellowships for young Ph.D.'s so they could acquire more training in their fields. Between 1924 and 1929, the foundation awarded 509 fellowships in the fields of physics, biology, chemistry, and mathematics, in that order, and the recipients included Werner Heisenberg, Enrico Fermi, and J. Robert Oppenheimer.[10] The foundation was supported by its own investment portfolio, which ebbed and flowed with the market, but even in lean years the sums available were substantial by the standards of the time. During the 1930's, in the depths of the Depression, the foundation's division of natural sciences disbursed about 1.5 million dollars annually, which represented "a substantial fraction" of the funding for basic research in the United States.[11]

The Rockefeller Foundation also helped create the field of molecular biology. In January 1933, the newly appointed director of the natural sciences division of the Rockefeller Foundation, Warren Weaver, completed a draft of the comprehensive funding plan that he would present to the trustees in April.[12] The major thrust of Weaver's proposal was that "Our understanding and control of natural forces has outrun our understanding and control of animate forces," so that the foundation should fund those sciences, such as biology and psychology, that offer the possibilities of such control, as well as those areas of physics, chemistry, and mathematics that are fundamental to advances in the life sciences.[13] By April, these critical disciplines had been further narrowed to "'four closely correlated and overlapping subfields,'—namely, endocrinology, genetics, 'biology of reproduction, and psychobiology.'"[14] Weaver called this program "Vital Processes" to distinguish it from traditional biology and to transcend what the foundation referred to as the "artificial" boundaries between fields imposed by academic departments. The director of the Rockefeller Foundation, Weaver's boss, was strongly supportive of this interdisciplinary effort, and in his own presentation to the board of trustees he underscored the complementary nature of the organization's support for biological and social science: "The Social Sciences [Division], for example, will concern themselves with the rationalization of social control.... The Medical and Natural Sciences will, through psychiatry and psychobiology, have strong common interest in the problems of mental disease."[15] Weaver's proposal was tantamount to a comprehensive definition of the modern symbolic body, for it was based not only on faith in science as the primary agent of social and moral transformation, an ideal promulgated by progressive creeds since the middle of the 17th century, but, more importantly, it gave these ideas a distinctly contemporary twist by proposing to implement them through a practical program of scientific control of biological processes, using physics as the lever arm.

Significantly, Warren Weaver was not a biologist but a professor of physics at the University of Wisconsin; and he accepted the position at the Rockefeller Foundation in part because mainstream physics had largely passed him by. In the late 1920's and early '30's, quantum mechanics became the dominant paradigm in physics; and Weaver, in spite of his lifelong fondness for the mathematics of probability, never accepted it as a physical theory.[16] Instead, in 1932, he took up an administrative post at the Rockefeller, where he set about implementing his quantitative vision of life founded on the principles of physics.[17] He prepared for his new duties by wading into biology for the first time, starting with genetics, then working his way through textbooks in cell physiology, organic chemistry, developmental biology, and so on, one after the other, until he felt conversant with the issues. Not long after, Weaver coined the term "molecular biology," and subsequent funding by the Rockefeller Foundation reflects this new vision of biology as a protégé of physics.[18] In 1935, the year when Weaver's program hit its stride, the foundation awarded grants to researchers who worked in "electron microscopy, in the use of radioactive and 'heavy' isotopes as tracers to probe biological tissues, in the application of X-ray crystallography and spectrographic analysis to organic material, in the development of more reliable ultracentrifuges and the analytic technique of electrophoresis."[19] To Weaver, physics and biology were united by their common goal of bringing nature, including human nature, under rational control, which is why he could, in a single year (1941), funnel money into both Florey's research on penicillin and Lawrence's cyclotron at Berkeley.[20]

Although there are differences in content and emphasis between physics and biology, they are united in their underlying rationale, for both have as their stated objective the rational control of natural processes, as well as a similar covert atheological agenda, a similar sci-fi projection, a similar vision of history as disembodied brains, a similar reliance on the Galilean circuit, and a similar relation to the state. The differences there are between physics and biology function primarily to maintain the structural opposition of the two disciplines within the technocratic cosmology. Although both physics and biology act on the biological body, they act in different ways; and although both are structurally related to the theory of progressive evolution, they occupy opposite ends of the arrow of linear time. Biology uses physical techniques, such as radioactive tracers and chemical analysis, to make physical images of biological bodies and to control biological processes, whereas physical science, through its invention of nuclear weapons and more conventional munitions, is primarily concerned with the killing and burning of biological bodies.

Also, physics and biology are responsible for different portions of the cosmic time line. Physics accounts for the creation of the universe, the evolution of stars and solar systems, and the gloomy heat death of everything. Biology, however, is responsible only for the last fractional percentage point of cosmic time: the counter-entropic bubbles that form in the wake of mid-sized stars. This evolutionary theory of the universe is an expression of the relative social status of biology and physics within the technocratic system of thought. Anthropologists point out that linear time is usually a code word for hierarchical social relationships. If a culture arranges technological innovations in linear time, as x → y → z, then this arrangement asserts the superiority of x over z or z over x. Generally, the oldest is the most primitive, as in diagrams of organic evolution and histories of technology, and to the technocrat it is usually the most inferior as well; but sometimes it is better for things to be old, as with art, antiques, or family fortunes. Generally, in modern American culture, the relationship of rank order to time depends on which side of the culture/nature dichotomy one is on. If the item is unequivocally high culture compared to its opposite, then older is superior. If it is unequivocally nature, then older is primitive and inferior. This is why physics is the older but still superior science. Physics is culture by definition, whereas biology is a "young science," inferior because it is more contaminated by blood and more closely connected to the earth; moreover, the "nature" that biology studies is a late emergent, a cooled accretion of starlight left over from the Big Bang. Because the subject matter of biology is doomed to be crushed in universal gravitational collapse or fried by a bloated sun, the discipline can never have the power, relevance, or generality of physics. To the technocrat, matter is primal and life but derivative.

To symbolic anthropology, however, the different branches of science play complementary roles in the construction and mutilation of the symbolic body, so whenever a piece of nature is removed by one branch of science, its man-made replacement is constructed by a different, complementary branch. Whether or not one judges penicillin to be good and A-bombs to be bad, these developments play complementary roles in the underlying symbolic body, irrespective of the morality of their use. Moreover, the technocracy is never so naive as to create an opposition between different branches of science that could be construed by the public as a simple polarity between good and evil. The essence of the one-two punch is the self-sustaining relationship between mutilation and prosthesis; and, for this reason, physical science not only destroys biological bodies with nuclear weapons, but it also conspicuously reconstructs them, providing X-rays, radiation therapy, and artificial hearts and limbs. Conversely, academic biology kills as much as it cures, for vivisection, dissection, and "animal models" are the standard tools of the trade.

Physics and biology play complementary but hierarchically differentiated roles within the symbolic body. Physics provides the ultimate mechanistic model of the universe, whereas biology brings the organic world into conformity with the assumptions of physics by demonstrating that the soft, wet, interdependent web of organic processes "is really" a hard, dry, crystalline, structure explicable by energy transactions. Conversely, to the extent that biology is successful in reducing life to matter, so physics must elevate matter to life. Biology and physics are complementary processes of mutilation and prosthesis, joined in a mutual agenda of transforming life on earth into images of celestial fire under the control of a male apparatus. Nowhere is this imagery better exemplified than in the Manhattan Project—the making of the first atomic bomb.

The Image of Vitality: Man Tanned and Hatted

The Manhattan Project exemplifies mythical and ritual processes in four main ways.[21] First, the atomic bomb has its historical antecedents in the imagery of radiation, not in physical theory, and it exemplifies the symbolic body as defined by the myths of Nebuchadnezzar's Dream, Stumps & Stars, and The Lone Galileo. Although the overt mission of the Manhattan Project is to make the weapon to end all wars, the covert agenda is to create a ritual artifact that best exemplifies the underlying mythology. Second, funding agencies, not physicists, provided the rationale for the development of nuclear physics during the 1920's and '30's, and they saw physics and biology as complementary endeavors, intimately related in a larger cosmological framework of progress and rationality, not simply as technological research and development. Third, the Manhattan Project is itself a ritual that incorporates the participating scientists into the postwar symbolic body, defining their subsequent social roles and academic status. Indeed, all of the major participants in the atomic bomb project ended up as heads of institutes, advisors to governments, and presidents of scientific societies. Fourth, the artifact created by the ritual is projected onto the landscape as an objective description of natural events, and its imagery becomes enshrined as a scientific theory of the universe, as exemplified by such theoretical notions as black holes, The Big Bang, and the like.

The anthropological hypothesis is that the detailed architecture of the atomic bomb reflects the mythology and social organization of the society that built it; and the dominant myth is the creation of a purified male body, separated by intellectual brilliance from an inferior and polluting nature conceptualized as female. Moreover, the male body thus created is hierarchically ranked, as in Nebuchadnezzar's Dream itself, into higher and lower elements, symbolized by head and tail. Although scientists when interviewed by journalists never fail to mention collegial relationships and the free flow of ideas, science is in practice a profoundly hierarchical institution, in which the participants see themselves as an intellectual elite, set apart from ordinary mortals by the timeless brilliance of their minds. The anthropologist Sharon Traweek confirms this in a study of contemporary particle physicists at the Stanford Linear Accelerator (SLAC). She shows that individual physicists rank themselves in a shared status system based on their putative achievements in physics, and that SLAC, far from being collegial, is organized into competing teams of scientists. Indeed, all of the particle-physics research institutions in the world are linearly ranked in a single status hierarchy.[22] At SLAC, different research groups do not interact socially; in fact, it is rare when their members even know one another; and in more than a year of fieldwork the anthropologist was the only person seen moving regularly across group boundaries. Moreover, Traweek shows that physicists acquire their social status through a series of exclusionary cuts, analogous to the cuts on Galileo's body discussed in Chapter 4, that eliminate would-be physicists at each level of professional advancement.[23] Although these ranking differences are supposedly based on scientific achievements, "pure science," as Traweek points out, cannot be separated from the control and display of genital imagery. Particle physicists work with detectors, experimental devices that record the collisions of subatomic particles, and in Traweek's words: "The language used by physicists about and around detectors is genital: the imagery of the names SPEAR, SLAC, and PEP is clear, as is the reference to the beam as 'up' or 'down.' One must see the magnets at LASS to appreciate the labial associations in the detector's name, Large Aperture Solenoid Spectrometer. Ironically, the denial of human agency in the construction of science coexists with the imaging of science as male and nature as female. Detectors are the site of their coupling...."[24] Also, as Brian Easlea, himself a physicist, observes in *Fathering the Unthinkable:* "In what one might take to be the austerely asexual world of theoretical and experimental physics practised at Los Alamos, it is surely remarkable how both sexual and birth metaphors frequently make an appearance in the writings, memoirs, and reminiscences of the participants."[25]

In the thin air of the Los Alamos mesa, the rarified social organization of academic physics, augmented by the enforced isolation, the exclusion of women, and the rigid status hierarchy imposed by the military, becomes embodied in the ritual objects that we know as atomic bombs. To avoid the charge of biased interpretation of the historical events surrounding the development of atomic weapons, I base my argument on facts drawn only from canonical, technocratic sources. A canonical source is operationally defined as a book that meets most of the following criteria: (1) It received an enthusiastic review in the *New York Times Book Review*. (2) It was supported by a grant or fellowship from a major American foundation, such as Ford or MacArthur. (3) It acknowledges an intellectual debt to a professor at Harvard University. (4) It won a major award for scholarly or literary excellence. (5) It is published by a major New York publishing company. (6) Portions of it were serialized in the *New Yorker* magazine. And (7), the author is interviewed on a respectable TV show such as *The World of Ideas With Bill Moyers*. When a book meets all of these criteria one can be sure that it is more orthodox than the pope, a Perfect Seven, and for this reason I have drawn primarily on the recently published book by Richard Rhodes, *The Making of the Atomic Bomb*. Rhodes begins by acknowledging his debt to Harvard biologist E. O. Wilson (Exhibit 7.1) as well as to both the Ford and Alfred P. Sloan Foundations. The book was published by Simon and Schuster and lauded by the *New York Times*. It also won the National Book Award, for it is well researched, well argued, and well written, with a meticulous attention to documentation. However, Rhodes's interpretation of the Manhattan Project reinforces the premises of the technocratic world view, presenting the atomic bomb as a logical development from scientific theory—while ignoring the entire social, symbolic, and philosophical context that makes its premises seem rational and self-evident. In particular he fails to recognize that the Manhattan Project was a cult of male solidarity, that its rationale was atheological not technological, and that its product was a ritual object that defined the major categories of the postwar symbolic body.

In addition to researching canonical, technocratic, literary sources, I have visited both Los Alamos and the neighborhood of the Trinity test site, taking notes and photographs of the facilities and museum exhibits to supplement the published accounts. These materials indicate that the imagery of the Manhattan Project has all the features that one would expect on the basis of the charter mythology as described in previous chapters of this book. Although rationalized in terms of weapons development, its deeper motivation was the purification of the male body by means of a man-made phallic apparatus and the transformation of celestial light into images of power.

The Man-Making Process

The Lone Galileo is first and foremost a purified male body, in which the contaminated parts have been cut away and replaced with a man-made Y-forked organ—a radiant head in phallic and cephalic aspects. Thus, the Manhattan Project is an implementation of the same man-making agenda that Weaver hoped to achieve with genetics, and this explains the otherwise curious choice of the primary Manhattan Project site, Los Alamos, a wilderness boarding school that offered well-to-do parents the promise of turning their boys into men. The Los Alamos site, or Site Y (!), as it was officially called, did not meet the original engineering specifications of a site for the proposed research laboratory which General Groves had given to his aide Major Dudley in the autumn of 1942. The original specification had requested a location west of the Mississippi River, at least two hundred miles from an international border, remote from prying eyes, with some existing facilities, and room for 265 people. Most importantly, the site should be shaped like a bowl with hills ringing the perimeter, supposedly to make it easy to string barbed wire.[26] Dudley located two possible sites that met these criteria: the town of Oak City, Utah, in the desert west of the Wasatch Range, and Jemez Springs, New Mexico, on the western slope of the Jemez Mountains not far from what is now the town of Los Alamos. Dudley rejected the Oak City site, his first choice, because too many people would have to be evicted, so he recommended Jemez Springs and took Oppenheimer there for an inspection. The scientific director did not like the site, finding the canyon too confining, and he suggested instead that they consider the boy's school on the top of the mesa, which

Dudley had examined and rejected because it was not the bowl that Groves had requested. Oppenheimer and Groves drove immediately to the Los Alamos Ranch School, and they approved it on the spot. Although it was not technically a bowl, it was an inverted one, a mesa, formed by erosion on the slope of the biggest bowl of all, an extinct volcano fifty miles across, the Jemez Caldera, visible from the moon. Not long after, the director of the Ranch School received a letter from the United States Secretary of War, Henry Stimson, seizing the property and evicting its inhabitants.[27]

Oppenheimer's recommendation for Site Y fits the imagery of the man-making process perfectly, and this interpretation is reinforced by the extensive space devoted to the Ranch School in the Los Alamos Historical Museum, which reconstructs portions of the original school. The school specialized in "making men" through a program of Old West wilderness skills, and it boasted the first mounted Boy Scout troop in the United States. Students slept on bunks in log cabins, and each was assigned his own horse to groom. In brown uniforms and Scout regalia, they rode in military formation like the Seventh Cavalry, camping out in the rugged back country that surrounded the mesa—imagery that is almost exactly recreated in the opening scenes of the movie *Indiana Jones and The Last Crusade*.[28] Moreover, the Jemez Mountains were associated with a similar phase in Oppenheimer's own life. As a sheltered, sickly youth of seventeen, raised in Manhattan, he was sent out West in the summer of 1922, in the care of his English teacher, to get toughened up. He stayed at the Los Pinos[29] dude ranch in New Mexico, and he went on a pack trip to the Jemez Mountains, on horseback, camping out along the way. The trip began in the Cañon de los Frijoles, which is now in Bandelier National Monument, ascended the Pajarito Plateau, and traversed the broad, grassy plain within the bowl of the Jemez Caldera. As Rhodes expresses it: "From an ill and perhaps hypochondriac boy he weathered across a vigorous summer to a physically confident young man. He arrived at Harvard tanned and fit, his body at least in shape."[30]

The Secret of the Atomic Bomb

Once the site for the Manhattan Project was chosen, its construction and staffing unfolded with incredible secrecy. Scientists disappeared from their university positions without any explanation, and any mail for them or their families was sent to a single post office box in Santa Fe, New Mexico, where Army censors read every letter.[31] FBI agents swarmed over potential project members, collecting thick dossiers on friends and associates as well as on the principals themselves. The mesa, dubbed The Hill by its residents, was surrounded by barbed wire fences, guard dogs, and mounted patrols of military police. Visiting scientists were assigned pseudonyms, such as Nick Baker for Niels Bohr, and trips to Santa Fe, only thirty-five miles away, were difficult and infrequent. The scientists at least had the satisfaction of their work, but their wives and families were at best passive spectators, as the project or its objectives could not be discussed at home. Judging from the published memoirs, most wives indeed remained ignorant of the goal of the Manhattan Project until after the press coverage of Hiroshima.

This security is always presented in terms of military secrets and the danger from spies, but this cannot be the whole explanation. White-helmeted MP's riding patrol on the isolated mesas of New Mexico must have prompted more speculation than they muted, and the columns of trucks carrying supplies to a high-tech enclave in the wilderness would certainly have piqued the professional curiosity of any local German spies. Presumably, however, there were no Nazi agents living in rural New Mexico, for the only documented spy who ever turned up, Klaus Fuchs, was brought there by the project and given a security clearance by the U.S. Army itself.[32] Moreover, the "secret of the atomic bomb" is not something that is easily stolen. That the Americans were supposedly building the atomic bomb to prevent the Nazis from getting it first, attributes to the Germans prior knowledge of the device, and it was European physicists, after all, many of them German, who explored nuclear fission in the first place. Moreover, any reader of the *New York Times*, *Collier's* magazine, or *Harper's* could have learned in 1939/1940 that uranium could be used to produce a bomb powerful enough to destroy New York; that the rare isotope of uranium, U-235, needed to be separated by physical or chemical means from its more common companion, U-238, in order to get bomb-grade uranium; and that a separation method based on the differences in weight between the two isotopes was theoretically possible.[33]

Clearly, it was not the *possibility* of the bomb that was being kept secret, rather the fact that the Americans were building it. But if this was "the secret of the atomic bomb," then it was betrayed by the very security procedures that were instituted to protect it, for the Russians, at least, learned of American intentions quite early in the war when all the prominent nuclear physicists in the United States disappeared from the scientific literature. Perhaps "the secret of the atomic bomb" was the exact method that the Americans were using to build it. This would indeed have been useful to an enemy, but the Americans built two different kinds of bombs, a plutonium bomb and a uranium bomb, with two different methods for achieving critical mass, and with fission material obtained by three different methods of production—because the Americans did not know "the secret of the atomic bomb" either. Moreover, even if one walked away with a safe full of plans, the production of the atom bomb still required an investment of two billion dollars (at 1945 purchasing power) and the creation of an industrial complex equal in size and complexity to the American automobile industry—all built within two years.[34] So how does one steal this? Clearly, the security precautions of the Manhattan Project were primarily instituted to keep the scientists in and the American public out, converting the exclusive but still civilian community of nuclear physicists into a full-fledged secret cult, whose content was known only to the initiated, and from which women and children were pointedly excluded.[35]

Metaphors of Life

In the 1930's the frontier of physics was defined by the process of irradiating heavy elements such as uranium and radium with beams of radiant energy and displaying the fragments thus produced by means of electronic counters and photographic plates. When subjected to this hammering, some elements became radioactive, that is, began exposing photographic film or triggering counters on their own, whereas others changed into substances lower or higher on the periodic table that were recognizable to chemists using chemical techniques. For example, uranium, element 92, at the high end of the periodic table, when bombarded with neutrons, can fission into the elements barium-56 and krypton-36, or tellurium-52 and zirconium-40, as well as other possible combinations.[36]

In the histories of physics, these interactions are seen as the consequence of "new discoveries"— nuclear fission and the transmutation of elements, and this terminology is reinforced by a folk theory of science that rationalizes changes in dogma as dictated by nature herself. Anthropology, however, defines "a scientific discovery" as *a ritual process that produces an image of nature that reinforces the charter mythology*. Because the charter mythology is the real (but covert) referent in scientific research, the new image does not have to follow logically from textbook dogma, and it may even contradict it. This anthropological interpretation resolves a perennial problem in the history and philosophy of science, namely, that some scientific "discoveries" can shake science to its foundations even when they contradict well-established dogma, whereas others, equally empirical, are dismissed or ignored *because* they contradict well-established dogma. However, if the real agenda in science is the imagery of the control of nature, not the explanation of nature, then more powerful images take precedence over theory. Moreover, it follows that discoveries are only bolts from the blue when they are interpreted within the framework of scientific theory itself, which, as shown in previous chapters, is a post facto rationalization that systematically excludes from consciousness the operative mythical elements, such as the purified male body and the celestialization of the hero. When viewed from this larger, mythical framework, however, most fortuitous "scientific discoveries" turn out to be *predictable.*

Thus, in the 1930's, the underlying ideology of physics *demanded* that the atom be split, for science needs to show that the limits imposed by nature are always subordinate to technique. Science also had to create elements where there were none before—the transuranium elements—because "nature" is always the initial stage in a process of domestication that replaces images of wildness by man-made prostheses that simulate natural processes. The method of discovery is predictable too, for science is a process of purification that separates earth from sky through control of radiant energy by means of a phallic apparatus. Given this symbolic context, it is hardly surprising that physicists in the 1930's focused on uranium, the element that defined the upper limit of the periodic table, and that the finest scientists in the world felt compelled to hit it with all they had, pelting it with particles, until—guess what—it broke! And if it had not broken? No problem—for scientists would simply recognize that uranium was not the vehicle of the sky god after all: The theory had been "wrong" and the element misnamed. Then the textbook writers, like the historians in Orwell's Ministry of Truth, would forget that uranium had ever existed and rewrite history until it led logically and inexorably up to whatever scientists were engaged in now. After all, the periodic table had been created by physicists in the first place, as had the dogma of the unbreakable atom and the names of the elements: What science giveth, science also taketh away.

It follows that splitting the atom is not a discovery but an agenda, which in physics was denoted by the technical term atomic *fission.* This term was borrowed from cell biology, where it denotes the splitting of a cell into daughter cells. Spencer Weart describes how atomic fission was given its name "at Christmastime," 1938: "The uranium nucleus, quivering and elongating until it broke into two pieces, reminded him [Otto Frisch] of the mysterious central transformation of birth, the division of a living cell. A biologist friend told him what it was called: fission."[37] Of course, no one had ever seen a uranium nucleus quivering, for this is an imaginary rendering of a mathematical construction; but biological metaphors, however contradictory to the mechanistic dogma of physics, far from being an idiosyncrasy of Frisch, also occur in the terminology of radioactivity, where the length of time it takes for one radioactive substance to decay into another is measured in terms of "half-life." Moreover, the word *decay* itself is biologically charged, normally denoting a loss of power or vitality or a change from living to dead. Also, the radioactive elements appear to be vital, glowing with an unearthly blue light, and the man-made metal plutonium even feels warm to the touch, in the words of one scientist "like a rabbit." These vital associations hint at the relationship between physics and biology discussed above, for fission is being equated with asexual reproduction, and radioactivity is conceptualized as a process of "decay," whose products are "half alive." These associations are extended and enhanced when the technical term *decay product* is replaced by the more informal *daughter element*, which in effect associates women with death and dangerous substances.

The concept of a chain reaction is also a biological metaphor, for it mimics one of the essential properties of life, namely, self-reproduction. Both the man-made element plutonium (Pu-239) and a rare isotope of uranium, U-235 (which accounts for less than one percent of naturally occurring uranium), behave similarly when bombarded by neutrons: They fission, and they give off more neutrons than they take in. Scientists speak of "generations" of neutrons being created in such reactions.[38] Moreover, physicists quickly concluded that if these substances were assembled in sufficient quantity, a quantity called the *critical mass*, then neutron bombardment could induce what came to be called a *self-sustaining chain reaction*, in which the breakdown products from one fissioning atom would set off fission in neighboring atoms, releasing enormous amounts of energy, while requiring only a small jolt to set it off. In short, the possibility of a nuclear chain reaction not only held the promise of a significant new weapon but it vividly reaffirmed the underlying structural role of physics as the science that vitalizes matter.

Although first "discovered" by physicists in the 1930's, the idea of the chain reaction, as well as its relationship to an atomic bomb, had already been formulated two decades earlier in science fiction.[39] In *The World Set Free*, published in 1914—one of Wells's forgotten "political" tracts that the anthologists of science fiction say hold no interest for a modern audience—H. G. Wells presents a complete ideology of nuclear weapons development, sketching both its military potential and the vision of future universal peace that would follow from its development. In this fantasy of the future, "atomic bombs" destroyed whole cities, but the war was followed by a united world government and an era of universal peace, in which atomic power was used to water deserts and melt the arctic wastes. Thus, the concept of the atomic bomb, far from developing fortuitously from discoveries in pure science, *preceded* most research in atomic physics, and its potential political implications drove the scientific agenda. Moreover, in the initial

Exhibit 5.1 The uranium agenda

In *The World Set Free* by H. G. Wells, published in 1914, the complete ritual cycle of atomic energy is presented: from the transformation of uranium into a man-made element to the development of "atomic bombs." As shown by the following quotation, the act of dropping the bomb is graphically sexual.

"His companion, a less imaginative type, sat with his legs spread wide over the long, coffin-shaped box which contained in its compartments the three atomic bombs, the new bombs that would continue to explode indefinitely and which no one had ever seen in action."[41]

"The gaunt face hardened to grimness, and with both hands the bomb-thrower lifted the big atomic bomb from the box and steadied it against the side. It was a black sphere two feet in diameter. Between its handles was a little celluloid stud, and to this he bent his head until his lips touched it. Then he had to bite in order to let the air in upon the inducive."[42]

vision of the atomic bomb, it is no ordinary weapon but a vehicle for social transformation and universal peace. In short, the atomic bomb, from the first glimmering of its possibility, was seen as intrinsically eschatological, which is why hundreds of physicists, some of them self-proclaimed pacifists, could be so easily persuaded to put aside their indifference to politics and fight the war to end all wars. In canonical interpretations of Wells's contribution, this sci-fi author "anticipated" science, but this way of speaking is just a ploy to maintain the autonomy of physics. Leo Szilard, the person who perhaps was most responsible for launching the Manhattan Project, admits that the idea for a nuclear chain reaction based on the fissioning of uranium by neutrons came to him after reading *The World Set Free.*[40] Science and science fiction are two aspects of the same process of mythic implementation. In the terminology of anthropology, science fiction creates the myth and science does the ritual.

In the same way, radioactivity was seen as a vehicle of eschatological transformation almost from the moment of its discovery at the turn of the century. As Weart shows from a careful compilation of historical sources, radium was initially hailed as the elixir of life and fed to the sick as medicine, so that any historical account of the atomic bomb that traces its development primarily to theoretical advances in particle physics while ignoring the parallel applications in medicine is seriously at variance with the historical facts.[43] The technological mimicry of life is fundamental to the mythology of the industrial age; and the physics of the 1930's did not so much "discover" the self-sustaining chain reaction as happen upon a vehicle that could finally give it physical expression. Significantly, Rhodes's account of the naming of plutonium confirms these associations between nuclear physics and vital processes in an unconscious but very revealing way. The physicist Glenn T. Seaborg, in Rhodes's words, "would name element 94 for Pluto, the ninth planet outward from the sun, discovered in 1930 and named for the Greek god of the underworld, a god of the earth's fertility but also the god of the dead: *plutonium.*"[44] But Pluto is a Roman god, not Greek, and, while associated with the underworld, he is not a god of fertility but of wealth. The confounding of life with death, wealth with fertility, and culture with empire is definitive of the modern technocrat.[45]

The Control of Pollution

In the anthropological interpretation of the atomic bomb, the elements uranium and plutonium have very different symbolic status. In the early 1940's, uranium was the *highest* naturally occurring element in the periodic table, the biggest atom of terrestrial matter; and although it had official celestial associations, being named for the planet Uranus, it also had more sinister connotations because it implicitly referenced the Greek sky god who was castrated by his son Cronus. Moreover, in modern English, the name Uranus, originally pronounced *yoo-ran'-us*, with a long "a" and the accent on the second syllable, is nowadays usually pronounced *yoor'-an-us* by newscasters and spokespersons, for the former sounds too much like the phrase "your anus" which conjures up the perineum and contradicts its celestial associations. Thus, uranium, as a radioactive element that occupied the terminal position in the periodic table, defined the "natural" limits of science, so it had to be broken and replaced by man-made elements. With the "discovery" of neptunium in 1940, made by bombarding uranium with neutrons, it became possible to reinterpret the periodic table, imposing a dividing line between "natural" and "man-made" elements.[46] Thus, uranium is suddenly demoted to the highest "naturally occurring" element, whereas neptunium plays a mediating role between the natural world of uranium and the unequivocally man-made world of the transuranium elements exemplified by plutonium. The anthropologist Edmund R. Leach points out that the English language likes to use binary contrasts in classifying biological phenomena, which then it mediates with a transitional category often fraught with emotion: for example, domestic animals/wild animals + pets; or females/males + homosexuals.[47] From this perspective, neptunium, named for the Roman god of the sea, is the mediating category between the natural and the unnatural elements—just as water mediates between the sky and land.

The Y-Forked Organ

In the imagery of purity and pollution, uranium, although far superior to lower elements, is nonetheless found in nature, and hence it is more terrestrial and natural than plutonium, which is a man-made element at the outer limits of the periodic table. Plutonium, moreover, is not just man-made but, more importantly, *vital*—for the stench of death is never far away. Seaborg, a native English speaker, in addition to naming the element *plutonium*, chose the abbreviation that would appear in the periodic table of the elements, namely *Pu*. In the American slang of the time the pronunciation of these two letters meant a fetid odor such as sewage, and Seaborg chose the initials, he said, "for the reason you might expect."[48] Since plutonium is an unambiguous man-made element, and by definition purified, the olfactory component must be due to its underworld associations and its relationship to vital processes, for in science *real* vitality comes from the sky, whereas terrestrial life stinks with the blood of females. Thus, the elements uranium and plutonium form a natural but contrasting pair on opposite sides of the neptunium boundary—a powerful but natural element (U) and a powerful but unnatural element (Pu). Moreover, the latter is associated with bad smells ("P. U.!") and the underworld (Pluto), suggestive of the power of death.

The associated imagery, described in detail below, indicates that purified uranium (U-235) is the distillation of a "natural power," and equivalent to a male principle, whereas plutonium is the simulation of an "unnatural power," albeit a very dangerous one. The theory of the symbolic body leads one to expect that these contrasting elements would be assimilated to the underlying schema of the Y-forked organ, with uranium symbolizing "male nature" (the lower phallic branch of the "Y") and plutonium the "man-made", transcendent head. Thus, there are two kinds of male power symbolized by elements 92 to 94: phallic power, exemplified by warfare and phallic display, and scientific power, characterized by the transformation of a polluting, "female" nature into celestial light. Thus, it is not surprising that the Manhattan Project produced not one type of atomic bomb but two, Little Boy and Fat Man, and that their architectures are in perfect accord with anthropological expectations. However, when atomic bombs are assembled, they look like any other bombs, steel cylinders tapered at both ends with stabilizing fins, and Fat Man is not noticeably different from Little Boy, just shorter and fatter, so the dramatic differences in design are not apparent in their external shape. To understand the symbolism, it is necessary to delve into the details of their construction.[49]

The type one bomb, a uranium gun, dubbed Little Boy by those who worked on the Manhattan Project, consisted of a long steel gun barrel with a target of purified uranium (U-235) at one end and a uranium bullet at the other. When the uranium bullet collided with the uranium target, critical mass was achieved.

The type two bomb, an implosion bomb, dubbed Fat Man by those who worked on the Manhattan Project, consisted of a nickel-plated plutonium sphere wrapped in gold leaf, about the size of a grapefruit, which was surrounded by a cylinder of natural uranium and two concentric spheres of high explosive (Figure 5. 1). When the explosive detonated, it compressed the plutonium, thus achieving critical mass.

In addition to Little Boy and Fat Man, the two branches of the Y-forked organ are unequivocally displayed at Ground Zero, Trinity Site. The Fat Man bomb, the test device, was put on top of a tower many stories tall—the image of The Lone Galileo—and wherever there is a disembodied head, a disembodied phallus cannot be far behind.[50] Significantly, the scientists of the Manhattan Project also erected a second, smaller tower at Ground Zero, which had no technical function whatsoever, but from which they suspended a huge steel cylinder, weighing many tons, called Jumbo. The cylinder had been prepared as a containment device for hot plutonium in case of a fizzle, but there was no scientific mandate for naming it after a famous circus elephant, suspending it like a pendulous organ, and vaporizing it in the first atomic blast.[51]

The two branches of the "Y" are also represented by the two heads of the Manhattan Project, General Leslie Groves and Dr. J. Robert Oppenheimer.[52] Groves was a corpulent man, a compulsive candy eater, with a substantial stomach that hung over his belt. Oppenheimer, in contrast, was almost preternaturally skinny, over six feet tall but weighing less than 130 pounds. In photographs of the two of them together, Oppenheimer looks like a tall, gangly youth next to Groves's apparently short and stout physique. This image of Groves and Oppie as the fat man and the little boy is ubiquitous in histories of the period, and in the Bradbury Science Museum at Los Alamos it is rendered by life-size statues of the two men.[53] Thus, Groves represented the adult male head in civic and military aspects, Oppenheimer the phallic power of the young male organ. This latter power is, of course, dangerous and suspect because it penetrates women, where it can become contaminated by Red—and certainly the history of the postwar McCarthy period, when Oppenheimer's security clearance was revoked, suggests that he elicited these symbolic associations.

In the iconic deep structure of technocratic mythology, the two heads of the Y-forked organ are connected via heartwood, a male vegetative principle, so one would expect Fat Man and Little Boy to be joined together by the trunk of a tree. But surely the Manhattan Project, with its obsession with metals, would skimp on the arboreal associations? Not so. Although the official designation of the weapons laboratory was Site Y, it was commonly referred to as Los Alamos, Spanish for "cottonwoods," the species of tree that lined the canyons of the mesa.[54] Cottonwoods are very appropriate symbols, for they are in the willow family, Salicaceae; and the word *willow* is derived from the Proto-Indo-European root **wel-*, meaning "to twist" or "to wind," like the sun's path along the ecliptic. In its Greek form, the root **wel-* is the word for "sun," which gives us the English word *helix*. This association between sky and tree is fundamental to Manhattan Project imagery, as it was in pre-Christian European mythology. In the latter, certain species of trees were sacred to the sky god, such as oaks, which were associated with Thor's hammer (lightning, element 90), whereas other species, such as groves of poplars, were associated with the entrance to the underworld.[55] Not surprisingly, it was General *Groves* who commandeered 59,000 acres in Tennessee for the site of the world's largest factory, the U-235/238 separation plants at *Oak* Ridge. Also, every historical account of the Manhattan Project, no matter how cursory, tells us that the first atomic bomb (a Fat Man) was tested at Alamogordo, New Mexico. This perennial historical tidbit is revealing, for Alamogordo, which means "fat cottonwood" in Spanish, is not particularly close to the Trinity test site but is about sixty miles away, across formidable deserts and mountains, as a glance at any map will show. Also, preparations for the Trinity test were organized from Los Alamos, and the bomb components went directly from Los Alamos to Trinity without going anywhere near Alamogordo. The connection between Trinity and Alamogordo is purely semantic—the name hauled in by the ears to ensure continuity between the immature alamos of the Ranch School and the mature, "fat" alamo of the first successful nuclear test.[56]

The geographical extent of the Manhattan Project is symbolically important too, for it encompassed the entire 48 states, from the Rad Lab at Berkeley to Groves's office in the Pentagon, from Hanford in the Northwest to Oak Ridge (also called Clinton) in Tennessee.[57] The two-dimensional relationship of the sites is important for understanding the underlying symbolism. The uranium-235 for Little Boy is produced at Oak Ridge, almost due east of Los Alamos by Tennessee *Eastman,* a subsidiary of the Kodak company, whereas plutonium-239 for Fat Man is produced in "the Northwest," at Hanford, Washington, by the Du Pont Corporation. The Trinity site, in contrast, is almost directly "below" Los Alamos in a desert called Jornada del Muerto, "the Journey of the Dead." The ostensible purpose of these interconnected facilities is to build atomic bombs, but underlying the technical pathway is a process of symbolic transformation that encodes a complete mythical cycle. In the symbolic process, natural uranium (a mixture of rare U-235 and common U-238), originally obtained from the Congo, is conceptualized as a primal, vital substance. Using huge banks of electromagnetic separators, built of silver from Fort Knox, and millions of feet of pipe circulating through membranes, the rising sun (east man) removes from natural uranium the celestial element, namely pure uranium-235. This is sent to the Ranch School to be transformed into a phallic apparatus of explosive power. Meanwhile, at Hanford, natural uranium is irradiated and transformed, in huge, concrete containment buildings called Queen Marys, into an even more dangerous substance—plutonium.[58] These tanks are supposedly called Queen Marys because they were as big as the ocean liner of that name[59] but the other Queen Mary in Anglo-American folklore is *Bloody* Mary—Mary Queen of Scots, whose name fits the anthropological interpretation far better. The latter Mary was imprisoned all her life in a castle, then finally beheaded, a story which encodes both the restrictions on women and the image of the bloody stump. "Bloody" Mary also associates to menstrual blood, a taboo substance that is "dangerous" to males. After all, it is females that structurally define the androcentric cult in the first place; and it is the *daughter* elements produced by radioactive decay, and the elements "your anus" and Pu itself, that require men to put lead shields over their genitals and goggles over their eyes; it is the planets furthest from the sun that require storage in massive steel cylinders touched only by artificial hands.[60]

Thus, the "power" of the bomb is partly due to its incorporation of taboo substances. Whereas U-235 is a purified, natural substance that can be directly incorporated into a gun (Little Boy), plutonium is an underworld element with a bad smell ("P.U."), produced by bloody marys, that needs to be transformed into spheres; plated with a silvery, reflective skin; daubed with gold; and embedded in concentric layers of uranium and high explosive. Then, the bomb containing this underworld substance is taken "below" Los Alamos (200 miles below on a map), in a place called Jornado del Muerto, the journey of the dead, and fired off just before dawn. This blast, described as "brighter than a thousand suns," fills the underworld with the light of a new, man-made dawn—and regenerates the symbolic body from the disembodied head of the Fat Man.

The Levels of Progression

The head of the Fat Man (Figure 5.1) is itself a complex symbolic structure that replicates the world view of the society that built it. In order to create a critical mass, the plutonium sphere needs to be simultaneously crushed by a blast of high explosive and constrained by a mass of natural uranium. Thus, the plutonium sphere is first coated with nickel to prevent oxidation, and (in the case of the first bomb, at least) wrapped in gold foil to smooth out the bumps on the nickel skin. Then the plutonium hemisphere is surrounded by a layer of uranium "five feet thick" and two layers of high explosive. Finally, the whole sphere is encased in a steel shell.

The explosive layer has a complex internal architecture. Within Site Y there was a laboratory called the X division which developed a high explosive especially for the implosion bomb, composed of TNT plus some other materials. The high explosive, abbreviated HE in the literature, was formed into pentagonal lenses (*lenses* are the official name), shaped like pyramids with truncated points, that fit together at their bases to form a sphere, with the flattened points facing inward toward the uranium and plutonium. Although mottled blackish-brown in color, the HE pyramids are true lenses, focusing the explosive force inward, as optical lenses focus light, and changing the shape of the shock wave to achieve maximal implosion. Thus, the concentric metallic spheres of the inner circles are surrounded by two spheres formed of flat-topped pyramids, like stylized suns but with the rays facing inwards.

These concentric layers of Fat Man encode the social hierarchy, with the highest-ranking images at the center and the lowest at the periphery (Figure 5.1). At the center is the element from Washington (the state), named for the first president of the United States, the father of his country, an unequivocal male head in political/military aspect. This sphere is covered with nickel, which is not just a chemical element but also an aspect of American money, for a *nickel* is a five-cent piece, which in the 1940's would have commonly carried a picture of an Indian warrior's face on one side and a picture of a buffalo (American bison) on the other. This nickel has strong cultural associations to skin, for every American knows that buffalo were hunted for their hides and that Indians took scalps.[61] Also, the fact that a nickel is a 5-cent piece is important too, for it reflects the pentagons in the HE layers. (It is also not an accident that General Groves built the Pentagon before he built the atomic bomb.)

There is geographical symbolism too. The natural uranium for the Manhattan Project came from what was then the Belgian Congo, now Zaire, in central Africa. At the beginning of the war, over a thousand tons of pitchblende, a uranium ore, were shipped to New York by the Belgians to keep it from falling into German hands, and one of the first acts of the Manhattan Project was to buy it.[62] At the time, the only other significant source of pitchblende, and hence uranium, was Joachimstal, an ancient mining center on the German-Czech border, from which the Curies and other nuclear pioneers had obtained their raw materials. From the perspective of the 1940's, natural uranium is an Old World element, both primal and close to nature, that mixes both pure and impure states ("feet of both iron and of clay").

The importance of Zaire to technocratic imagery is affirmed by books such as *Congo,* a technological thriller written in 1980 by a card-carrying member of the technocratic establishment, Michael Crichton (Harvard education, M. D. degree, former fellow at the Salk Institute for Biological Studies).[63] In this book, an American expedition, controlled remotely via satellite from Houston, is sent to find a lost city in the African rain forest. Associated with the lost city are mines containing the world's only source of certain crystals with semiconductor properties (Type IIb boron-coated blue diamonds) that are essential for the next generation of high technology. The city is guarded by a man-made race of ferocious gorillas that crush the skulls of visitors with stone paddles (that is, half-men that implode hemispheres); and the Africans, of course, are portrayed as either faithful porters or as the equivalent of animals, eating human flesh and defecating on the seats of an airplane. Interracial imagery was no better forty years ago, and the core of the atomic bomb symbolically encodes the same geographical hierarchy, namely: Washington (power) → New World (money) → Old World Europe (high culture) → Africa (primitive beginnings).

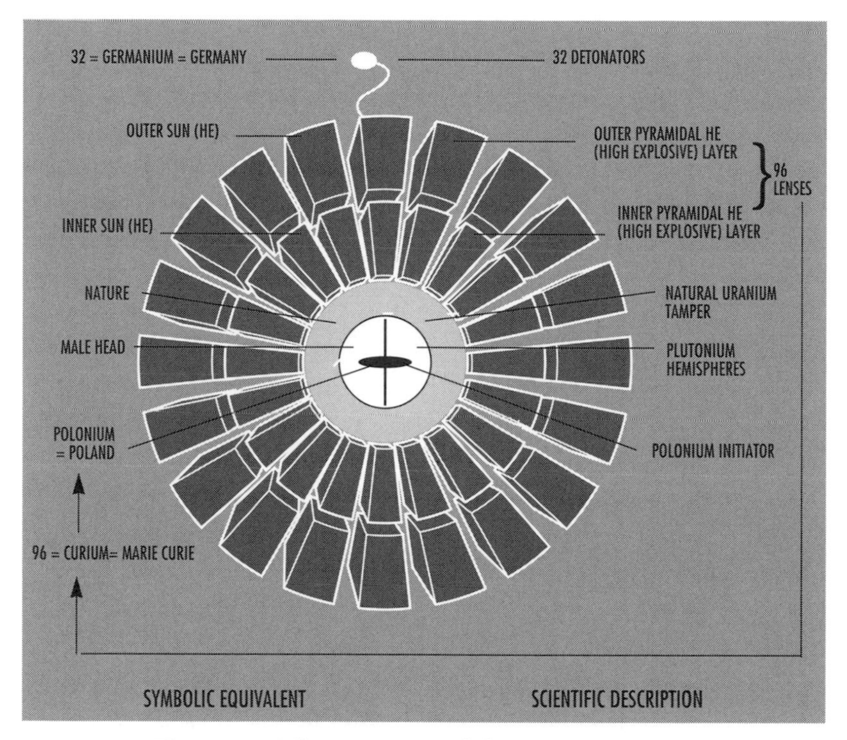

Figure 5.1 The structure of the Fat Man bomb encodes its symbolic associations.

In the Fat Man, the social hierarchy of Nebuchadnezzar's Dream interacts with a pair of near-simultaneous explosive events. For the bomb to detonate with maximal yield, the high explosive must be detonated to compress the plutonium hemispheres to a critical mass, and then, within ten-millionths of a second, a neutron source must be activated to initiate the chain reaction in the superdense plutonium.[64] Although initiator design is classified, Rhodes says that the neutron source was probably the radioactive isotope polonium-210, a decay product of radium, which was named by Madame Curie for Poland, her native country. The polonium initiator was probably buried inside the plutonium sphere, so that the high-explosive blast would strip off its protective shield, releasing neutrons into the now critical plutonium and initiating the chain reaction. The high explosive in turn was triggered by conventional detonators wired to the outside of the HE sphere, 32 such detonators in the case of the Trinity bomb. These two critical events, implosion and initiation, are as much symbolic as they are technical, for their names and numbers encode a capsule history of the Second

World War. To a physicist, 32 is the atomic number of germanium, an element that is named after Germany, as polonium is named after Poland.[65] Thus, in the architecture of the atomic bomb, a "germanium" blast engulfs polonium, initiating a chain reaction, just as the German invasion of Poland started the Second World War.

In addition to the hierarchical male body encoded in the head of the Fat Man, there is an explicitly biological metaphor in the outer, high-explosive layers which illustrate once again the congruency of physics and biology. Although the entire body of Fat Man is the product of Site Y, which is also the name of the male chromosome, a portion of this spherical complex is produced by X division, a part of Y, which is also the name of the female chromosome. In this respect, the Fat Man resembles a complete diploid cell, with both male and female *nuclear* material (Ah!). In this androcentric inversion, the female portion is on the outside, looking in as it were, and the contribution of X division is the ugly, mottled, black and brown spheres of the TNT mixture, which contrast unfavorably to the gleaming metals at the core. After all, the high explosive in the A-bomb is abbreviated "HE," which is the chemical symbol for the element helium, which in turn is derived from the name for the Greek sun god. The HE mixture has an interesting symbolic structure of its own, for the pair of concentric high-explosive spheres are formed of truncated pyramids and shaped like imploding suns. This pair of inverted suns, the inner and outer layer together, we will call the Double Helix. Although the pair of suns take the form of concentric spheres in the plutonium bomb, their symbolic structure is actually more complex than this, for *helix,* as mentioned above, is derived from the Proto-Indo-European root **wel-,* "to twist" or "to turn," like the seasonal progression of the sun along the path of the ecliptic. So the Double Helix of the Fat Man can also assume the morphology of a pair of symmetrical twisted strands—the Watson-Crick model of DNA.

With the advent of DNA, the vitalization of matter by physics and the mechanization of life by biology went into high gear, the two disciplines feeding on each other until they reached critical mass. A year after Edward Teller vaporized the island of Elugelab with the first hydrogen bomb, spilling radioactive fallout across the Pacific,[66] Watson-Crick developed their model of DNA. In the technocratic state and its institutions of higher education, there is of course no relation between these two events because one is biology, the other physics, one progressive, the other morally flawed, but symbolically they are complementary images, drawing on the same iconic deep structure. The A-bomb is a fission device modeled on asexual reproduction and implemented by splitting the nuclei of heavy elements such as uranium or plutonium, but the H-bomb is a *fusion* device, implemented by merging of the nuclei of light elements such as hydrogen or helium. Thus, where the first phase of atomic weapons development implemented a bacterial model of vitalization, namely, the splitting of a nucleus, the second phase emphasized a sexual model, namely, the fusion of two separate nuclei into one. Although H-bomb design is still classified, Rhodes illustrates a possible architecture, which combines a spherical fission bomb with a new invention shaped like nested cylinders. In fact, the architecture of the H-bomb is a good test of the anthropological theory precisely because it has never been made public but should nonetheless be predictable on purely symbolic grounds. If the theory presented in this book is correct, then what Oppenheimer called a "technically sweet" solution to the problem of building a fusion bomb should reflect the imagery of sexual as opposed to asexual reproduction: that is, an architecture in which separate male and female organisms exchange nuclear material by coupling.[67] Since it is against the law to publish classified information, even if created in the mind's eye, we will refrain from further speculation,[68] noting only that the gender of the offspring of this union has long been public and never in doubt. When the first successful hydrogen bomb, named Mike, was exploded in the South Pacific, Teller himself sent the word to Los Alamos: "It's a boy."[69]

Exercise 5.1. If the twins Hiroshima and Nagasaki were symbolized by a space flight, what would be the launch date, given the following?

- Apollo 11, which made the first manned landing on the moon, was launched on July 16, 1969, the anniversary of the Trinity test.
- Apollo 17, the last lunar mission, was launched on December 7, 1972, the anniversary of Pearl Harbor.
- Gemini VII was launched December 4, 1965 and returned on December 18, after rendezvousing in space with Gemini VI.
- There are 1,317 days from Trinity to Pearl Harbor, and three days from Hiroshima to Nagasaki, August 9, 1945. The twins bring the war full circle (in reverse order!).

Because of their differences in structural role—with biology mechanizing life and physics vitalizing matter—the two sciences never engage in identical research, so in this respect they are indeed "distinct disciplines," but their activities are always coordinated by an identical infrastructure of myth and imagery. The structural complementarity between physics and biology, the one-two punch, is exemplified by the close historical relationship between the development of the atomic bomb and the discovery of penicillin. Although these events are associated in the popular imagination with two radically different aspects of science, one militarist and dangerous, the other medical and humanitarian, they are nonetheless united on the level of imagery, funding, application, and technique. Penicillin, although less dramatized and less overtly political, is as much a reflection of the unconscious symbolism of Big Science as are Fat Man and Little Boy.

Penicillin is a naturally occurring substance, fatal to many species of bacteria, that is produced by molds of the genus *Penicillium.* Molds are downy growths of fungus that appear on decaying organic matter, and they reproduce by means of microscopic spores that are wafted along by the wind. The discoverer of penicillin was a British physician, Alexander Fleming, known to his colleagues as Flem, who for many years had studied bacteria at his laboratory in St. Mary's Hospital in London. Many bacteria species cause diseases in humans, so they had long been known to the medical profession, for example, *Staphylococcus aureus*, which causes boils and carbuncles. Fleming grew colonies of staph so he could describe and analyze their properties; and, like other bacteriologists, he grew them on small glass dishes, called petri dishes, which are first smeared with a substance that is nutritious to microbes, "inoculated" with a small amount of the desired organism, and finally incubated at the appropriate temperature. Typically, the bacteria reproduce in such vast numbers that they form a film on the surface of the dish that is visible to the naked eye.[70]

In the traditional account of the discovery of penicillin, one fostered by Fleming himself in his speeches and publications, a spore of the fungus *Penicillium* had blown in through the window of his laboratory while he was on vacation in the summer of 1927, contaminating a petri dish of staphylococcus that he had left out. When he returned to work he observed that the bacteria in proximity to the mold growth had been killed, forming a telltale ring of dead microbes around the intruder.[71] In the mythology of science, purifying agents come from the sky, so it is not surprising that he claimed that the mold came in by the window; but Ronald Hare, an historian and bacteriologist who knew and respected Fleming, doubts this classical account. Hare has done experiments to show that the killing of bacteria by an accidental inoculation of the penicillium mold, far from being common and unremarkable, in fact can only take place under a very narrow range of temperature conditions and bacterial growth, so that if the weather had been different that summer, the antibiotic effect would never have been observed. Significantly, Fleming himself tried to "rediscover" penicillin on many occasions by leaving petri dishes exposed to airborne contamination but was never able to reproduce the original observation, with its tell-tale ring of dead bacteria. Hare has also argued that the spore was more likely to have come in by the door than the window: researchers who were studying the role of molds in asthma reactions had assembled, downstairs in the same building, mold specimens from all over London. Moreover, the story of Fleming's discovery of penicillin only became important long after the event, for dozens of antibacterial substances and organisms had been described in the scientific literature prior to 1927, all of which were eventually dismissed as lacking clinical significance; and penicillin too was ignored, even by its discoverer, for twelve years, until it was resurrected, in 1938, by a man named Howard Florey.[72]

Where Fleming was a scientist of the Victorian mold who worked in a cubicle on his own, supported in part by his own income, Florey was a thoroughly modern academic researcher who battled for grant money, assembled teams of co-workers, and kept his eye on the main chance. It was Florey who made penicillin scientifically respectable when, for reasons that are obscure to his biographers, he committed himself and his laboratory to a crash program on the antibacterial substance that Fleming had described more than a decade earlier.[73] Whatever Florey's motives, medical concerns were not high on the list, for as he himself says: "I don't think it ever crossed our minds about suffering humanity; this was an interesting scientific exercise. Because it was some use in medicine was very gratifying, but this was not the reason that we started working on it."[74] One of Florey's biographers suspects that these remarks may be only a cover-up for a genuine humanitarian concern; but if this be so, what does it say about scientific ideology when a Nobel laureate in medicine has to disguise humanitarian motives as dehumanized problem-solving?[75] Whatever his motives, in 1938, he made the decision to commit a research department at Oxford University to the investigation of antibacterial substances, including the untried penicillin. Since this decision does not follow logically from theoretical antecedents in biology or from Florey's own career, the mythology of the decision looms all the larger in the historical account, and there are two versions of the myth available. In one, attributed to Florey himself, he was standing under a chestnut tree at the moment of decision, in the other, he was passing under an elm discussing scientific problems with his biochemist Ernst Chain.[76]

In the theory presented in this book, trees in science represent heartwood, the shaft of the Y-forked organ, so an inspirational "chest nut" is no surprise. But elms? They are important because they have "winged" seed pods called samaras. The traditional emblem of the medical profession is the winged staff of Asclepius, an aspect of the god Mercury, the patron of physicians and thieves, around which is coiled a chthonic serpent; and the origin myth of Florey's decision encodes a condensed image of this symbol: a wooden vertical shaft plus a winged seedpod plus the image engendered by the name "Chain"—who was, the biographers imply, wrapped around Florey's little finger. In this context, the choice of penicillin, derived from a fungus named penicillium, a word that means "little penis," is not so lacking in precedent after all. Moreover, it is a good little penis that attacks the evil seed: *Staphylococcus* is derived from *coccus,* a bacterium of spherical shape, from Greek *kokkos,* a kernel, seed, or berry; and *staphylo-,* from Greek *staphyle,* a bunch of grapes.[77]

It is at this point in history, in September 1939 (three days after Hitler invaded Poland), that Florey, under the chestnut/elm, and aided by his faithful biochemist, Ernst Chain (a Jewish refugee who fled Germany on January 30, 1933, the day Hitler became chancellor),[78] begins to turn the pathology department at Oxford University into a factory for the still unproved penicillin.[79] In a letter to a British funding agency at this time, Florey says that the production and purification of large amounts of penicillin could easily be accomplished, even though he must have known that this had been the stumbling block that led Fleming to abandon research on the new substance in the first place, and he asserts its nontoxicity to animals even in large doses, citing for this conclusion Fleming's experiment with one mouse and one rabbit. As grantsmanship, however, his assertions about penicillin were entirely justified, for the Natural Science arm of the Rockefeller Foundation—not its Medical Sciences arm—began active support of the project, leading, two years later, to Florey's meeting with Weaver in London. What is it about penicillin that made it such an attractive candidate for R&D at this particular time, in advance of the evidence that could justify such a commitment? The answer is that physics and biology play complementary roles within the rituals of reduction, and that it is imagery, not facts, that animates the players. When physics sets out to produce a mushroom cloud, then biology is enlisted to produce a cloud of mushrooms.[80]

Which Florey did—producing spores by the billions. His assistants grew penicillium on specially designed trays, drained off the fluid, and subjected it to endless physical and chemical assaults to separate the active penicillin factor from the biochemical soup in which it was mixed. Macfarlane captures the atmosphere at Florey's Oxford laboratory: "The practical classroom became the inoculation department, where the washed and sterilized pots were charged with medium and then inoculated with mould spores by paint spray-guns. Everything had to be done under strict aseptic conditions to prevent contamination by bacteria, and the girls were dressed in sterile caps, gowns, and masks. The inoculated pots were then wheeled on trolleys to what had been the students' preparation room, now converted into a huge incubator kept at 24 degrees Celsius. After several days of growth, the penicillin-containing fluid was drawn off from beneath its mould by suction, using the special equipment designed and made by Heatley, and the pots recharged with medium. After filtration, the crude penicillin solutions went through Heatley's extraction contraption. The air was full of a mixture of fumes: amyl acetate, chloroform, ether. These dangerous liquids were pumped through temporary piping along corridors and up and down stair-wells. There was real danger to the health of everyone involved and a risk of fire or explosion that no one cared to contemplate. But by February 1941, despite several serious setbacks, including a sudden break in the ability of the mould to produce penicillin, Florey had available enough material to begin his first trials in human beings."[81]

The imagery of penicillin production is completely cognate with the contemporaneous Manhattan Project even though there are no chemical or physical reactions in common. In both cases, a school is turned into a factory—the primal metaphor of modern education; and a natural substance is extracted and purified (U-235, penicillin) from a natural underworld substance (the natural mixture of U-235/U-238, the penicillium mold), using massive, specially designed apparatus attended by specially purified individuals who are at risk from a hostile and polluting nature (radioactive contamination, bacterial contamination). In both cases, the product is a new and purified substance that kills the enemies of mankind. In fact, one of Florey's biographers, Ronald Hare, himself a scientist, catches the image exactly in the title of his book: *The Birth of Penicillin and the Disarming of Microbes*. But there are also differences between penicillin and the atomic bomb, as would be expected from the contrasting structural roles of biology and physics. In the Manhattan Project, the enemy is theoretically the Nazi, for which the Japanese come to be substituted; whereas with penicillin, the official enemy is staphylococci, but syphilis and gonorrhea will become its most memorable victims. Also, in the case of physics, the final product is celestial fire kindled by purified, disembodied men, with women forbidden to be present, whereas in biology the product is a purified extract of nature, manufactured by masked women under the control of men, which purifies an already sick body.

In 1941, the British, now actively at war, and with clear clinical evidence of the efficacy of penicillin, tried to drum up financial interest in the new antibiotic on the other side of the Atlantic, and American drug companies, Squibb and the Merck Corporation among them, had in fact been experimenting with penicillin for a year or more. After the Japanese attack on Pearl Harbor, the United States government, through the Office of Scientific Research and Development, instituted a crash program of its own, infusing massive amounts of money and industrial organization into the growth and purification of molds, which transformed the production of penicillin from a factory into an industry. Florey shipped his assistant Heatley, along with samples of penicillium, to a government laboratory in Peoria, Illinois; and by D-Day more than twenty factories were engaged in its production. In this push to industrialization, the molds, of course, were irradiated to develop more productive, mutant strains; and millions of dollars were expended in a futile attempt to produce a completely synthetic version.[82] These expenditures were seen as facilitating the war effort by saving wounded soldiers so they could live to fight again. Thus, neither the development nor the industrialization of the drug had much to do with humanitarian motives. Rather, the job of biology is to show, as emphasized above, that the soft, wet, interdependent web of organic processes "is really" a hard, dry, crystalline structure explicable by energy transactions. From this point of view, Florey earned his Nobel prize in the summer of 1943, when penicillin was purified to the point that it could be transformed into crystals—not just any kind of crystal, but, as shown in Figure 5.2, crystals that look like exploding stars.[83]

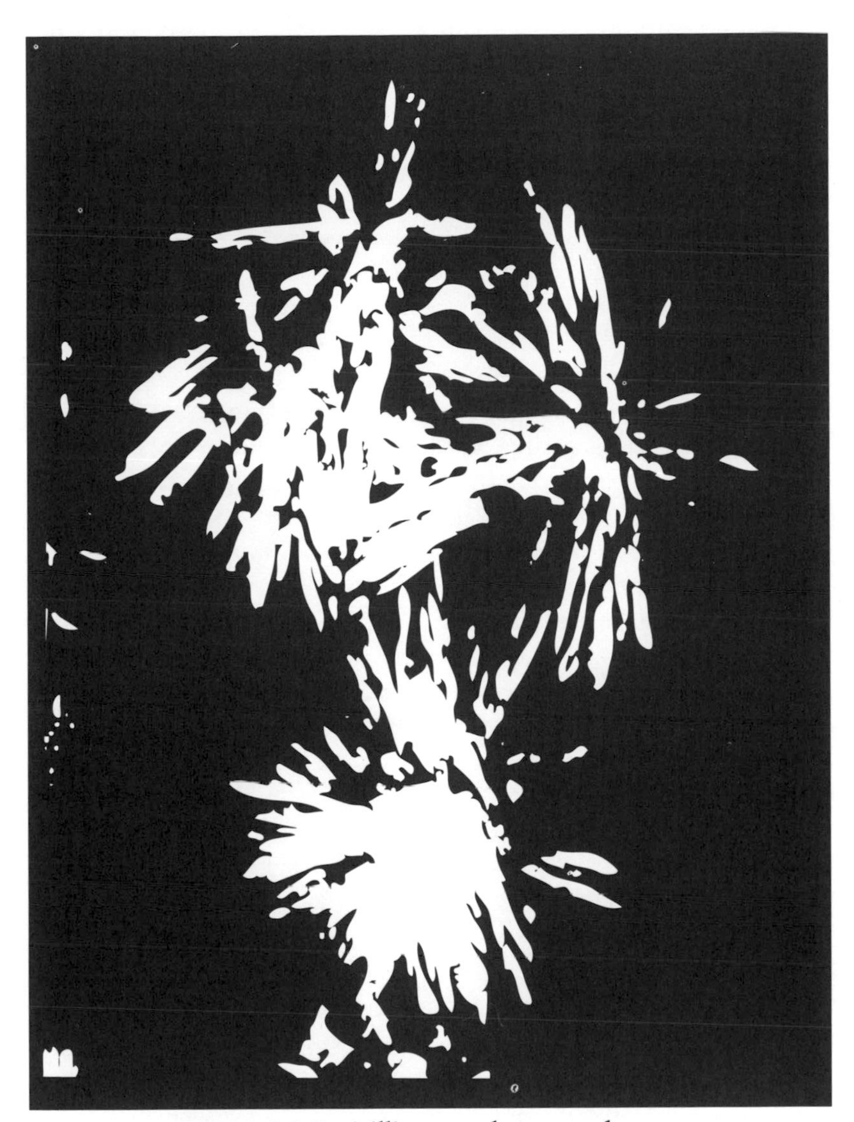

Figure 5.2 Penicillin crystals are starbursts.

The atomic bomb and penicillin are both productions of Anglo-American society at the identical moment in history, so one would expect great similarities in the underlying cultural assumptions; but symbolic anthropology takes this observation a step further, showing that both of these technological innovations tap a deeper level of unconscious imagery and myth, as shown in Figure 5.3. In this interpretation, science is a ritual process for transforming nature into culture, but structurally it occupies what anthropologists call a liminal zone, a transitional state between ordinary life and the changed status conferred by the ritual itself. This technological development process (Figure 5.3) parallels the purification of the body shown in Figure 3.4, but instead of transforming the body of the scientist into a disembodied brain, which is then transformed into a star by subsequent acts of brilliance, the ritual creates physical analogs of the celestialized body: ritual artifacts that express the imagery of purification, androcentricity, and disembodiment. In this way, even nonscientists and women can be incorporated into the purification process as participants. Also, Jews have a special role in this system as ritual experts. The origin myths of both penicillin and the atomic bomb attribute key discoveries to Jewish refugees from Hitler, and this is shown in Figure 5.3 as the incorporation of ritually pure Jewish brains into the brainchild—a peculiar feature of World War II mythology whose function will become apparent in the next chapter.

As shown by the feedback loops, the rituals of purification form a closed system in which the products of one scientific process become the raw materials for the next, but for convenience Figure 5.3 divides this closed circle into discrete steps, each of which is given a mnemonic name, such as the Mary Factor or Heartwood. These deep structure images are present in both penicillin and the atomic bomb, but they are exemplified in different ways. The phallic head, for example, is manifested by Little Boy in the Manhattan Project, but by the hypodermic syringe in the case of penicillin. The Underworld is represented by the Jornada del Muerto in the Manhattan Project, but by infected rats in Florey's clinical trials. Moreover, as names are an essential component of the symbolic process, deep structure concepts may be manifested by either their imagery or by their names or both, such as Florey's assistant Chain, who was both a Jewish refugee and the source of the "Chain" reaction that made penicillin possible. Because the system is a closed loop, each step in the ritual process creates additional substances that are themselves assimilated to the iconic deep structure. For example, exploding stars purify nature by means of fire, but they also create a residuum of ash which becomes the clay feet of Nebuchadnezzar and a residuum of star dust which is symbolically equivalent to airborne seeds.

Because the body of the New Man requires both terrestrial raw material and unequivocally celestial elements that are transmitted by prior acts of purification, the disciplines of biology and physics play complementary roles in the symbolic body of industrialism, responsible for blood and fire respectively. Thus, the modern symbolic body is a

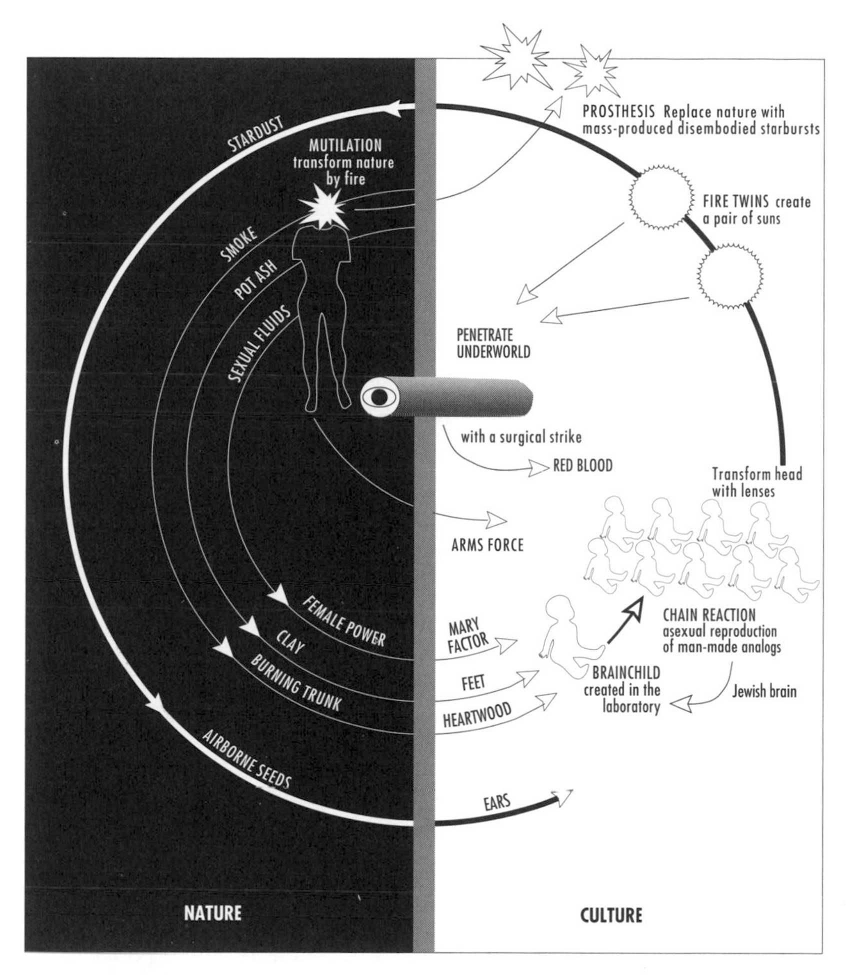

Figure 5.3 Penicillin and the atomic bomb are spin-offs of the ritual purification of contaminated substances.

composite of both biological and physical images, both of which are represented in the architecture of the atomic bomb and in the development of penicillin—and American moviemakers understand this relationship even if academic historians do not. In the 1950's Hollywood produced countless films that were predicated on a relationship between radioactive contamination and the development of huge, mutant animals known in the trade as "bug-eyed monsters."[84] "Bug eyes," as biologists have shown, are quite different in structure from vertebrate eyes. Instead of a single lens in the center which projects light on to the retina, "bugs" typically have a compound eye, consisting of multiple lenses arrayed on the surface of a sphere. This image of the bug's eye matches the spherical form of the high explosive layers of Fat Man, which, as we have seen, are formed into *lenses* whose bases form the surface of a sphere when viewed from the outside. The details of the construction of the HE layers make oblique reference to the imagery of blood as well as fire. TNT is clearly fire, but in the Fat Man that was exploded at Trinity, there were 96 high explosive lenses; and 96 is an element, curium, named by Seaborg after Pierre and Marie in 1946—in short, a fusion of a peter and a mary (bloody). This imagery, however far-fetched it may seem to literal-minded mechanists, encapsulates the history of 20th-century genetics (the study of "blood relations") in a deep way, for modern population genetics begins with Muller's experimental irradiation of fruit flies, *Drosophila,* and tracing mutations in the morphology of their eyes.

Thus, X Division's contribution to the structure of the Fat Man appears as a pair of suns when viewed from its own center, that is, from the sphere of Pluto (wealth and power), but it forms the head of a bug-eyed monster when viewed from a normal human perspective. The modern symbolic body presents two faces to the world: a bug-eyed monster that represents the outsider aspect of the HE spheres, and a pair of suns that represent the insider aspect. Moreover, the suns have alternative morphological forms, depending on whether it is physics or biology that is emphasized: sometimes they are shown as a pair of spheres, but at other times they are represented by their paths along the ecliptic, that is, by the sinuous molecule of the Double Helix.

The mechanization of life and the vitalization of matter both reference biological concepts, but neither of them has very much to do with normal biological processes, with sexual reproduction, or with biological gender differences. Plutonium, in spite of its warm, blue glow, is not the elixir of life but the most poisonous substance known; and X Division at Los Alamos, in spite of being structurally "female," was staffed exclusively by biological males with presumably normal XY chromosome complements. And molecular biology, far from being the intellectual revolution it claims to be, is the assimilation of life on earth to the imagery of the atomic bomb.

THE LAST DAYS OF MARCUS KARENIN

Hard Bodies/ Soft Bombs[1]

Vomiting it back up will be much more painful than swallowing it.

Nicole Sault[2]

Casablanca[3]

A major theme in the historical story of the atomic bomb is the flight of Jewish scientists from Nazi Germany to freedom in England and the United States: Einstein, Szilard, Von Neumann, Teller, and many more. As technocratic histories typically ignore the social context of scientific innovation, treating the latter as an autonomous development from pure research, it is especially significant that Rhodes devotes a whole chapter, called Exodus, to the stories of the Jewish refugees; and it ends in a familiar paean to freedom.[4] Also, in J. Bronowski's *The Ascent of Man*, the dawn of the atomic age begins with refugees leaving Germany and ends with a visit to Auschwitz. Thus, in modern mythology, the story of the atomic bomb is inextricably interwoven with the German persecution of the Jews and their rescue by the Americans.

Once again, the Rockefeller Foundation was a major behind-the-scenes protagonist. In 1933, shortly after Hitler became chancellor, Warren Weaver visited Germany, where he noted the government's role in the public burning of books by Jewish authors and confirmed that Jewish recipients of Rockefeller grants had already been removed from university payrolls. Back in New York, the Foundation decided to assist in the relocation of scientists from the Continent who wished to immigrate to England and the United States, and by the end of the year it had found jobs for seventy-one deposed academics. Appropriations for this purpose continued until the beginning of the war, by which time 197 refugees had been relocated at a cost of $750,000.[5] The Nazis, as everyone knows, advocated a political program of racial purification, but as Arno J. Mayer, a Jewish historian at Brandeis University, has demonstrated, the Nazi doctrine was not simply anti-Semitic but based on a deeper ideology of Blut und Boden (blood and soil), in which the body politic would be cleansed of its polluting elements, not only of Jews but of mental defectives and victims of genetic diseases as well. Significantly, as Mayer shows, the policies of purification of the body that culminated in the Holocaust were first implemented by the medical profession, not by storm troopers, as programs for the sterilization of the mentally retarded and as euthanasia for the mentally ill.[6] Moreover, the first extermination programs were directed against ethnic Germans, in German hospitals, not against Jews or foreigners. These developments are inexplicable from the conventional theory of fascism, in which fascists are equated with Nazis and Nazis reduced to anti-Semites, but they make perfect sense from the perspective of the symbolic body, which postulates an intrinsic relationship between the scientific world view and the symbolism of purity and pollution.

From this point of view, the Nazi is not some inexplicable aberration but a logical development from a system of mythical thought that was just as widespread in the United States as it was in Germany. But where the Nazis implemented the androcentric cult of purification and pollution through the selective breeding of a genetically superior race and the physical elimination of impure genetic strains, the British and Americans embarked on a complementary strategy, one quite different in its overt methods but quite similar in its underlying goals: namely, the creation of a purified race from scratch, using the scientific control of genetic material. That is, where the Nazi program of purification focused on the cutting off of the impure elements, the American plan emphasized the development of prosthetic devices to replace the impure genes entirely.

This interpretation explains the seamless transition in American science from eugenics to molecular biology. Until the late 1920's, the Rockefeller Foundation provided substantial funding for the field of eugenics, a social and political program that advocated the selective breeding of people in order to improve the genetic well-being of the species. Far from being a socially peripheral movement, eugenics had substantial support among the educated classes in the United States in the first two decades of this century, for genetic analysis almost always revealed the most perfect specimens of humanity to be well-to-do white males with degrees from East Coast universities; and it had been consistently funded by Carnegie money since the turn of the century, supplemented by substantial grants from the Rockefeller and Harriman families. Moreover, the biological laboratory at Cold Spring Harbor, which was to play such a leading role in the development of molecular genetics after World War II, had been established with eugenic policies in mind. In the 1920's, experts from the laboratory lobbied the United States Congress to help pass new immigration laws that gave priority to the so-called Nordic racial type, and contributed expertise to state-mandated sterilization programs for such genetically flawed individuals as alcoholics, epileptics, and criminals.

The eugenicists were forthright in their social and political aims, but by the 1930's intellectual culture had become thoroughly technocratic in spirit, dictating that self-serving political ends be formulated in the dispassionate language of science by certified experts who never advocated policy but merely called attention to inevitable developments. Also, with the rise of overt racism in German-speaking Europe, the times apparently dictated a more temperate and subtle approach. For example, Hermann J. Muller, an eminent geneticist who would go on to become a Nobel laureate in the field, denounced eugenics as scientifically flawed at a meeting of the American Eugenics Society in New York in 1932.[7] Although Muller himself believed in the selective breeding of superior humans, as a radical Marxist he saw the purpose of genetic reform as the creation of the Socialist New Man, not as the reinforcement of an existing class structure;[8] and years later he argued that this could only be achieved by means of artificial insemination, man-made genes, and implanted embryos—in order to circumvent the political objections of those deemed genetically inferior.[9] In 1934 the Carnegie Institution eased out its director, Charles B. Davenport, an outspoken enthusiast of eugenics policies, and within five years it had terminated all support for the field. About the same time, the Rockefeller Foundation also had second thoughts. With the ascent of a new director in 1928, the foundation began raising questions about the validity of eugenics programs, and by 1932 it had ceased supporting anything that called itself by that name. However, the replacement of eugenics by something more marketable must have been high on the trustees' agenda when, in the spring of 1933, Weaver and his boss presented their proposal for transforming human nature by means of the scientific control of genetic mechanisms. As Weaver himself expressed it in a memo to the committee: "Can we develop so sound and extensive a genetics that we can hope to breed, in the future, superior men?"[10]

The disproportionate role of World War II in American popular consciousness fifty years after the fact is due to the protective coloration provided by what I call the myth of The Nazi/ the Jew/ and the Yank. In both the Nazi and technocratic mythologies, the Jew is a symbolically charged character, credited with disproportionate power; but where Hitler dismissed physics as "the Jewish science," the technocracy sees the Jew as the driving force behind the atomic bomb. Where the Nazi symbolic system focused on Jewish blood, the technocracy focused on Jewish brains. Roland Barthes, in a paper called "The Brain of Einstein," has called attention to the role of the disembodied brain in modern mythology, where it substitutes for the person (see Exhibit 6.4), and Einstein contributed to this imagery by bequeathing his brain to science.[11] But as Einstein's brain is invariably paired with the migration of the Jewish scientists from Europe, it is not just a "genius" brain but a "Jewish" brain as well, emblematic of both the American rescue of the Jews and the leadership of American science. Thus, the Jew has a major symbolic role in the ideological systems of both Nazi Germany and the American technocracy, but these roles are opposite in sign. To the Nazi the Jew is a powerful negative force that can contaminate the body politic far beyond what demography would predict, but to the technocrat the Jew is a positive (but still foreign) force that when incorporated in small amounts into the symbolic body ensures intellectual potency.[12]

The rescue of the Jews by the Americans is told and retold by American Jews themselves, as in Herman Wouk's *The Winds of War*, a best-selling novel and a popular television miniseries; but this myth is nonetheless a serious distortion of history, for the vast majority of Jews were never rescued by anybody. As David S. Wyman shows in *The Abandonment of the Jews*, there was extensive, overt anti-Semitism in the United States in the 1930's and early 1940's, including parades by uniformed Nazi sympathizers and rallies in Madison Square Garden; and during World War II the country was all but closed to foreign immigration.[13] The War Refugee Board headquartered in the United States managed to move about 220,000 people beyond the reach of the Axis powers, often by placing them in camps outside the war zone, but it received 90 percent of its funds from Jewish volunteer organizations, not from the United States government. That the latter, in fact, did more to obstruct the rescue of Jewish refugees than to facilitate it, is borne out by the immigration statistics. In the four-year period between Pearl Harbor and Victory in Europe Day, no more than 21,000 refugees (of all religious persuasions) were admitted to the United States—a miniscule number even when compared to prewar immigration quotas. In the same vein, the Rockefeller

Foundation saw its mission as the rescue of Jewish *brains,* and it was completely indifferent to the vast body of European Jews, who were left to struggle as best they could. Thus, the rescue of the Jews, while admittedly beneficial to the handful who made it to England and America, had less to do with humanitarian motives than with the contributions that the survivors could make to the technocratic agenda—contributions that were not merely technical but symbolic as well.

The Rockefeller Foundation's rescue of Jewish intellectuals from Nazi eugenicists, which began in mid-1933, was perfectly synchronized with Weaver's program for the transformation of the gene pool using the techniques of physical science; and it was funded and administered by the same group of people. But the myth of The Nazi/ the Jew/ and the Yank creates the impression that the genetic agenda was confined to the other side of the Atlantic. Thus, the myth inverts fascist symbolism on a surface structure level while leaving the underlying schema of racial purification and Jewish potency unchanged. By equating fascism with anti-Semitism, the origin myth of the atomic bomb makes it logically impossible for Americans to be fascists, for did not the Yanks save the Jews and win the war against Hitler? The myth also reinforces the American image of themselves as the Chosen People, chosen by history as Jews have been chosen by God. Then, by projecting fascism into a distant and mythical past of jack-booted storm troopers, it makes these tendencies as comfortably remote from contemporary American life as Attila the Hun.[14] Finally, the story of the Nazi and the Jew focuses on something called the Holocaust, literally a "consuming fire," that happened long, long ago, in a galaxy far, far away. Thus, it follows that the immolation of Hiroshima and Nagasaki cannot be holocausts—for holocausts are done by Germans. By equating racist programs of extermination with overt selective breeding of the sort favored by the Nazi regime, the myth insulates both nuclear war and molecular biology from any totalitarian stigma.

Symbolic analysis of the myth of The Nazi/ the Jew/ and the Yank not only reveals discrepancies with America's historical role in Jewish immigration, but it demonstrates that the surface structure of the myth is the polar opposite of its deep structure content. The myth, while apparently pro-Jewish, is actually based on the transformation of anti-Semitic feelings into a ritual of purification that attempts to assimilate the innate power of the unclean Jew and to turn it to technocratic purposes. This transformation of the Jew from negative force to positive symbol can be explained by the dynamics of repression. When negative feelings toward an ethnic group are repressed, the result is not positive feelings but negative feelings with more elaborate disguises. For example, if the negative feeling is expressed by the proposition "Jews are greedy pigs," then a moderate level of repression can attenuate this proposition to "Smelly pigs [Jews] are Big Noses." But if this level of negativity is itself repressed without coming to terms with the original motivation, then the following transformation of content takes place:

"The Jew is a Big Nose." Taboo! →

"The Jew—lots of nose." Taboo! →

"The Jew nose a lot." Better! →

"The Jew knows a lot." Good!

Ergo: "The Jew is a Big Brain."

However, this surface structure content is antithetical to the deep structure symbolism that motivates it, so it engenders a new, more powerful, compensatory motivation, namely:

"Cut off his gnosis!"

But as this wish to mutilate the brainy Jew is even more unacceptable than the original symbolism of the greedy pig, and as it requires an overt act of violence, it manifests itself as its surface structure opposite, namely, the disembodied Jewish Brain purified by fire. It is this symbolically purified and domesticated Jew—not the ethnic Jew—that is incorporated into the symbolic body as Einstein's Brain. Moreover, as the repressed act of mutilation is unresolvable within the system, it is projected on to an outsider—the Jew's perennial nemesis, the Hollywood Nazi.

The putative potency of the purified Jewish brain is the reason that a Jewish scientist with a history of communist flirtations, J. Robert Oppenheimer, in violation of all the unwritten rules of the American officer corps of the 1940's, was catapulted over the heads of corn-fed physicists like Lawrence and put in charge of the United States Army's single most expensive, prestigious, and secret project. It is for this reason too that the United States Navy's only Jewish admiral was put in charge of developing nuclear submarines.[15] None of this, of course, can be overtly acknowledged by an equal opportunity employer, so it is marked by informal customs. *J. Robert*, an excised name with a prominent *J*, was known to his friends and colleagues as "Oppie," and he wore as a personal trademark a narrow-brimmed hat called a porkpie hat, which figures in Los Alamos foklore. That is, Oppie wears "pork" on his head, and *porkpie* with *rk* removed is an anagram of "Oppie." As the charter mythology of the A-bomb makes clear, Jewish brains in proper channels can be transformed into nuclear power.

The attribution of symbolic power to the Jewish brain is an essential component of technocratic mythology, dating back to the image of the New Jerusalem that inspired both the British colonization of North America and the millennial expectations of Nebuchadnezzar's Dream. In the imaginary future age of science and rationality envisioned in the 17th century by Francis Bacon, the new technocratic government is installed in "Soloman's House," a reference to both the biblical King Solomon, renowned for his wisdom, and to the first Jewish temple in Jerusalem, which was built during his reign.[16] By symbolically incorporating Jewish brains from the Old World, the modern technocracy not only appropriates the mythical power of biblical prophecy, but it creates a secular analog to the Protestant concept of the difference between Old and New Testaments: namely, an obsolete system of taboos transformed into a mechanism for personal salvation. Indeed, the atheological agenda of science requires this. If science is a mythical inversion of Judeo-Christian theology, and if Judaism and Christianity are related as Old Testament and New Testament respectively, then the inversion of this system is a pseudo-Judaism transformed by a pseudo-Incarnation into a pseudo-Christianity. As Lévi-Strauss might phrase it: as Jew is to Christian, so Jewish Brain is to the New Man. For this reason, Jews are symbolically necessary to the technocratic system—but only as an apocalyptic remnant purified by fire, distilled into gray matter,[17] and transformed into nuclear power.

This interpretation is supported by the term *holocaust* itself, which by any measure is an odd way to characterize the Jewish experience in the Second World War. Although *holocaust* literally means "completely consumed by fire," it is a Greek word, not Hebrew, and it is derived from Christian theology where it denotes the sacrifice of the purified animal that removes the sin of the community. In the Judeo-Christian context, the rite of sacrifice is not a fertility cult, as in Frazer (see below), but an expiation for sin, chartered by the Whole Offering or Holocaust of the ancient Jews, a rite described at length in Leviticus. Thus, the use of the term Holocaust by American Jews to describe the Nazi experience is itself an oblique reference to the technocratic myth of The Nazi/ the Jew/ and the Yank, for it implicitly makes Jews into sacrificial animals, admittedly male domestic animals without blemish, but at best a bullock, a ram, or a dove, the species mandated for the Whole Offering. In other words, the body of European Jews performs the same function in the dominant myth of the technocracy that the male domestic animal performed in ancient Judaism and that Christ performed for Christians. The Jew is the sacrificial victim who expiates the sin of the community.

What sin, then, did the community commit that demanded this level of sacrifice? Oppenheimer said to Truman, "We have known sin"—so the sin must be the atomic bomb. And who built the bomb? The Jews built the bomb—everyone knows that! Although they were specifically brought to the United States by the Rockefeller Foundation, organized by the U. S. Army, and given their marching orders by impeccable members of the Protestant establishment such as Roosevelt and Stimson, the association has been hammered home for so long that the development of nuclear weapons is now inconceivable without Oppenheimer, Von Neumann, Teller, and Szilard. Thus, the symbolic role of the Jewish brain in technocratic ideology reinforces the imagery of Jewish guilt and purification by fire that created the "Holocaust" in the first place!

This symbolic analysis of the charter mythology indicates that the Nazi agenda of racial purification through warfare and selective breeding, far from being an alien development which was exterminated by Americans in 1945, was simply implemented in the United States and Great Britain by a different and seemingly more benign strategy. Since Nazi Germany was unabashedly hierarchical, militarist, and racist, the doctrine of racial purity could express itself in overt form and be pursued through negative eugenics, but the United States was officially a democratic, equal-opportunity country, where such elitist doctrines could not be publicly expressed by government officials. When the United States and Germany came to blows, there was even more reason for maintaining their ideological distinctiveness. Consequently, the doctrine of racial purification, while implicit in the androcentric cult, is necessarily more repressed in the United States and hence more effectively disguised—but it is certainly no less prevalent. Where in Germany the repression structure of the Nazis was translated into official government programs to purify the blood and create a master race, using the methods of extermination and selective breeding, in the English-speaking democracies the repression structure was translated into a public denial of any such intentions—even as biomedical research was unconsciously converted into a systematic program for the scientific control of human reproduction and the construction of artificial genes. Thus, it is not Germans, but our own Luke Skywalker in a black tunic and high leather boots who must be projected into the past.

Mercury (Hg) Gate[18]

The overt eugenics of the 1930's was transformed in the United States and England into biomedical research to "conquer genetic diseases" through artificial bloodlines; and this development, like the atomic bomb itself, is chartered in the writings of H. G. Wells. In his sci-fi epic *The World Set Free*, written before the First World War, H. G. Wells combines in a single plot the themes of the persecuted Jew, the sacrificial victim, the resurrected king, atomic war, the control of human reproduction, and the cure of genetic disease—all wrapped up in a worldwide empire under the control of the English and the Americans.[19] In this otherwise undistinguished work, cardboard characters are moved around an imagined 20th-century stage in response to epic events created by science. There is no central hero, only a succession of observers, who are scarcely more than disembodied names that serve as vehicles of narration.

At the beginning of the book, the protagonist is Professor Rufus, reminiscent of Soddy and Rutherford, who inspires youth to split the atom by providing visions of huge explosions and limitless energy. He is followed by Holsten the chemist, who moves from research on the "cold light" of fireflies to the creation of artificial elements in 1933. He invents the Holsten-Roberts engine, which produces cheap atomic power and transforms bismuth into gold as a by-product. Holsten is succeeded in the narrative by one Frederick Barnet, who gives an account of his life between his nineteenth and twenty-third birthdays, when he was a soldier in the war that consumed the middle decades of the 20th century.[20] Barnet's narrative is interrupted by the first atomic bomb attack, on Allied headquarters in Paris, as seen through the eyes of a nameless woman—the only nameless character in the book.[21] The woman then disappears from the narrative as suddenly as she arrives, consumed by a rush of "hot water" from the dismembered lower body of a general killed in the attack.[22] Meanwhile, Barnet, campaigning in the Low Countries, first experiences atomic war in an analogous manner when he is caught in the flood that ensues when German A-bombs burst the dikes, leaving him and his men stranded on a barge, barely alive on a sea of desolation. Then the scene shifts suddenly to the serenity of the Italian Alps, near Lake Maggiore, at a place called Brissago, where King Egbert is discussing the future of the world with his secretary, Professor Firmin, a social scientist. The king says that a "World Republic" must be formed with a monopoly on atomic bombs, but his proposed political reforms are in fact completely autocratic, initiated from the top, and democracy is explicitly dismissed as a foil for the ignorant masses who lack the scientific sophistication to understand what is good for them.

It is at Brissago, on June 4, 1959,[23] that the French ambassador to Washington,[24] a man named Leblanc ("the white"), in cooperation with the Americans and the farsighted King Egbert, "the young king of the most venerable kingdom of Europe,"[25] assembles the ninety-three scientists and heads of state who will found the postwar world.[26] The conference takes place under the auspices of the King of Italy, with Leblanc presiding and King Egbert, who has renounced his title, serving as president. The meeting is held in the open air, as if at a camp, with the heads of state dining at three long tables, with no order of precedence among them.[27] The participants at the conference agree to centralize the control of atomic bombs, abolish war, and form a World Republic.

Wells also charters the Cold War, for the architects of world peace do not reckon with the duplicity of the Slavs. The King of the Balkans, Ferdinand Charles, nicknamed the "Slavic fox," is secretly plotting to take over the world while claiming to be a supporter of the international accords.[28] At 5 A.M. on the morning of July 17, 1959 (the Trinity test was at 5:29 A.M. on July 16, 1945),[29] a Slavic plane carrying three atomic bombs (in Wells, atomic bombs always form a trinity) attempts to blow up the Brissago conference, but the malefactors are run to ground before they can carry out their bid for world domination. Then, to finish things off, the King of the Balkans and four of his companions are caught in the act of smuggling atomic bombs; and all five are shot and killed. Egbert orders Firmin, his secretary, to bury them.[30]

With the death and burial of the Slavic fox, human society enters "Chapter the Fourth: A New Phase." It begins in the aftermath of the atomic war, and it is described through the eyes of Barnet, who has returned from the Continent to the ancient cathedral town of Winchester. The day the declaration of peace from Brissago is received, the cathedral bells begin to toll. In the years that follow, a single worldwide monetary system is set up, the metric system adopted, and English becomes the international language. Everywhere, society is organized on a rational basis, and politics disappears as a feature of the human condition.[31]

The character of King Egbert, who resigns his title to become president of the World Republic, is henceforth referred to by the narrator as the "ex-king"—and this appellation has far more sinister connotations than may be apparent to a modern reader. In one scene King Egbert discusses with Firmin the book *The Golden Bough* by Sir James George Frazer, a well-known work in Victorian anthropology. This book, which Frazer himself described as a multivolume footnote to an obscure line in Virgil, tries to explain the Roman ritual of the priest-king of Nemi by reference to similar customs in both the ancient world and among contemporary "savages."[32] At Nemi, a lake outside of Rome sacred to Diana, the priest of the shrine is a murderer who has killed the previous priest, but he is safe from prosecution so long as he performs his sacerdotal duties within the confines of the sacred site. However, an escaped convict may become the priest of Nemi in turn by killing the reigning priest. Frazer explains this custom (as anthropologists of his day typically did) as a "survival" from a hypothetical and more primitive past, a time when the roles of priest and king were merged in a fertility cult. Each year a new king was selected to reign as the embodiment of the solar cycle, and then he was killed to ensure the return of the sun. King Egbert tells Firmin that "authentic kings" are "cut up" as in *The Golden Bough*, and he acknowledges their sacrificial role by noting that they "sprinkled the nations with Kingship."[33] Thus, an "ex-king" is a dead king: the sacrificial lamb who is offered up for the good of the community.

The World Republic, run by the ex-king, far from being the secular state that one might expect from one of the most famous expositors of science, is, as Wells himself tells us, a successor to the "world religion" of Christianity.[34] Moreover, it was made possible by the "genius of Marcus Karenin" who successfully countered the secularizing trends of the 19th century, for "he saw religion without hallucination, without superstitious reverence, as a common thing as necessary as food and air, as land and energy to the life of man and the well-being of the Republic.... He gave it clearer expression, rephrased it to the lights of the new dawn...."[35]

Although Marcus Karenin is not specifically said to be Jewish, he is, like the Jewish scientists who worked on the atomic bomb, associated with escape from persecution, for his father "escaped from Russia" and settled in London.[36] As Wells was writing before the Bolshevik Revolution, Karenin is presumably a Red, a Jew, or both. He is also a victim of an unspecified congenital disease that has left him stooped and crippled, a condition that can only be cured by a potentially fatal surgical procedure. In the last chapter, "Chapter the Fifth: The Last Days of Marcus Karenin," the founder of the new world religion discourses to an enthralled audience of brilliant scientists, who gather at his bedside at a medical research facility built high in the Himalayas, where he will soon be operated on by the surgeon Fowler.

On the night before his surgery (which leads to his death when a "blood clot" from the successful operation blocks his heart), Karenin, wrapped in artificial furs, talks with his female admirers about equality of the sexes. He tells the women that they must no longer think of themselves as women, for the category of woman only exists in relation to men.[37] Rather, "you have to learn to think of yourselves—for our sakes and your own sakes—in relation to the sun and stars. You have to cease to be our adventure, Rachel, and come with us upon our adventures...."[38] The conversation turns to heredity. The physician Fowler tells them of recent research that reveals how the sex of children is genetically determined:

> "It is still, so to speak, a mere laboratory triumph," said Fowler, "but tomorrow it will be practicable."[39]
>
> "You see," cried Karenin, turning a laughing face to Rachel and Edith, "while we have been theorising about men and women, here is science getting the power to end that dispute forever. If woman is too much for us, we'll reduce her to a minority.... These old bodies, these old animal limitations, all this earthly inheritance of gross inevitabilities, falls from the spirit of man like the shrivelled cocoon from an imago...this round planet is no longer chained to us like the ball of a galley slave."[40]

Then Karenin, the women, the doctors, and a nurse go out on to the terrace to watch the setting sun, where Karenin once again speaks in the voice of the world soul, in which the individual is absorbed into the spirit of Athens and of Florence. But as the sun dips below the horizon, we get a glimpse of the real source of his power, which only emerges after dark. The theosophical Karenin is suddenly transformed into the prince of darkness, threatening the sun with a passion found nowhere else in the book:

> And you, old sun, you sword of flame, searing these poor old eyes of Marcus for the last time of all, beware of me! You

> think I die—and indeed I am only taking off one more coat to get at you. I have threatened you for ten thousand years, and soon I warn you I shall be coming. When I am altogether stripped and my disguises thrown away.[41]

The Spirit of Los Alamos

Wells's image of a scientific Messiah enshrined on a mountaintop in Tibet explains aspects of atomic bomb mythology that are otherwise completely inexplicable. The house historians all attest that Oppenheimer upon viewing the explosion of the first atomic bomb at Trinity quoted the words of the *Bhagavad Gita:* "Now I am become death, the destroyer of worlds."[42] But exactly when he said this is not clear. Oppenheimer's brother Frank, who was at Trinity when the bomb exploded, later recalled: "And I wish I could remember what my brother said, but I can't—but I think we just said, 'It worked.' I think that's what we said, 'It worked.'"[43] By calling the first atomic bomb test "Trinity" and by pairing it with a quote from scripture, Oppenheimer transforms an ugly weapons development project into a spiritual undertaking. Not surprisingly, the choice of name is highly mystified in the historical accounts. In some versions, Oppenheimer has John Donne's sonnets open on his desk: "Batter my heart, three-person'd God ..." Oppenheimer, in fact, set this scene himself for future historians in a letter to Groves in 1962, where he says his reasons for choosing the name Trinity are unclear but that Donne's words, written just before his death, had been on his mind:

> As West and East
> In all flatt Maps—and I am one—are one.
> So death doth touch the Resurrection.[44]

To the technocrat, there are no metaphors, only facts, so the zero point where east and west become one must be Ground Zero at Trinity, which explains why American Indians and Hindu Indians can be mixed together in the mystification of the atomic bomb without apparent contradiction. But the real reason for choosing the name Trinity is far more prosaic than the story of John Donne and the Christian resurrection. An essential ingredient of the atomic bomb is TNT, the high explosive that compresses the plutonium, and the power of an A-bomb blast is measured in the tons of TNT that would have to be detonated in order to produce an equivalent explosion. Since the entire Manhattan Project and Oppenheimer's personal reputation hinged on a successful blast, in the months before the test TNT became a word fraught with anxiety, taking on symbolic status. The letters "TNT" stand for "trinitro-toluene," so to account for the origin of the name "Trinity," just *fission* TNT by chopping off *ro-toluene* and add a Y-forked organ—Presto!—God.

The symbolic body of industrialism feeds on the religious traditions of all cultures, and nowhere is there a better illustration than Oppenheimer's oft-quoted passage "I am become death, the destroyer of worlds." This is a quote from Hindu scripture, as popular historians never fail to mention, so the subject of the sentence must be God in the form of His avatar Vishnu, the second person of the Hindu trinity. If this be so, then Oppenheimer must have wanted the name "Trinity" to be taken in a *spiritual* sense, not just as some arbitrary code name, and the legion of historians and journalists who have repeated his remark must also have thought that the theological associations are particularly relevant. But how can this be? Is not the atomic bomb the epitome of a 20th-century technical enterprise, the product of a mechanistic cosmology from which even the clockwork God of Newton had been banished long ago? The juxtaposition of the theological Trinity with the atomic blast of the same name can only be explained by the myth of The Lone Galileo. In the story of Galileo's triumph and trial, the scientific hero successfully challenges Church doctrine while disavowing any theological intent, and this juxtaposition of physics and theology charters the technocratic world view. Thus, the A-bomb, like any other scientific advance, is both a physical and theological statement: an objective image of nature and a covert inversion of Judeo-Christian doctrine.

But Oppenheimer does not say: "The technocracy now has the power of God," for that would be an overt theological statement, and overt theology is taboo. So Oppenheimer *quotes* God—not directly but obliquely: not the Judeo-Christian God of a clean, progressive, technical civilization but the polytheistic God of what Americans regard as a dirty, backward, third-world nation, a sink of misery and superstition. It is the second person of the Hindu trinity, Vishnu, whose words are linked with the Trinity test.

The Hindu associations are also found in technocratic movies such as *Close Encounters of the Third Kind*.[45] In this Spielberg film, highly developed aliens from outer space choose to reveal themselves to earthlings, who are, of course, equated with people in the United States; and they land their huge, luminous ship on an isolated, flat-topped mountain in the American West, Devil's Tower, Wyoming, which is reminiscent of the mesa at Los Alamos. The United States government clamps down a vise of secrecy on the visit so that only highly placed scientists and aeronautical engineers, plus a few interlopers, make contact with the higher life forms. However, before these events take place, the scene shifts suddenly to India, where thousands of saffron-robed devotees are shown chanting in unison, obviously aware of some impending spiritual happening. Why does a movie of such blatant technocratic premises feel obliged to reference Hindu spirituality?[46]

The atheological agenda requires that religious imagery be incorporated into the symbolic body of industrialism; but as the charter myth of The Lone Galileo specifically excludes the home-grown religious traditions of Judaism and Christianity from the definition of spiritual power, the technocratic system has to appropriate culturally alien symbols of spirituality, no matter how incongruent they may otherwise be when contrasted to the official ideology of mechanism, materialism, and technical progress. The alien religious symbols that surface in technocratic products are not to be understood as religious symbols in themselves, that is, as evidence of an exotic but nonetheless genuine spirituality, for this would require an understanding and appreciation of non-Western cultures which is precluded by Nebuchadnezzar's Dream. In science and technology, the non-Western religious symbols are always ethnocentric images in disguise, covert references to specifically Judeo-Christian religious symbols that cannot otherwise be overtly referred to within a scientific context. That is, Oppenheimer was not quoting Vishnu because he was a practicing Hindu but because he needed to make the reference to the Christian Trinity even more transparent and precise. If he quotes Vishnu, it is because he wants the Trinity test to be understood as referring to the Judeo-Christian Trinity in general and to the second person of the Christian God in particular. Oppenheimer is not quoting God but articulating the technocracy's image of itself: "I—the redeemer—have become the destroyer of worlds."

But lest these oblique references be misunderstood, the actual referent, namely, Christian theology, is made completely overt in the more popular popularizations. William L. Laurence, science editor for the *New York Times,* was the only journalist allowed to witness the Trinity test and to fly on the mission to Nagasaki, and in the first published descriptions of these events he gives them the appropriate theological spin. He records that on the day after the Trinity test he encountered E. O. Lawrence:

> "What a day in history!" he exclaimed.
>
> "It was like being witness to the Second Coming of Christ!" I heard myself say.
>
> It then came to me that both "Oppie" and I, and likely many others in our group, had shared in a profound religious experience, having been witness to an event akin to supernatural.[47]

Although many mechanists are indifferent to spiritual questions, the technocratic players are generally committed atheists and anticlericals, people who are struggling with the demons of religion: they are either sons of clergymen like General Groves or party intellectuals like Oppenheimer, who was a graduate of the Ethical Culture School, an educational program based on the premise that the human species must assume the responsibility that was formerly the province of God.[48] Significantly, Oppenheimer dabbled intellectually with spiritual questions all of his life, supposedly learning Sanskrit to read the Hindu scriptures in the original language; but there is no evidence from the life he lived that spirituality, in either Hindu or Western terms, was for him anything but another esoteric text to be mastered. In this respect, Oppenheimer pioneered what I call the *Oppie Effect*, the pseudospirituality of the late 20th-century technocracy, in which scientific theories take on the properties of God.[49] Oppenheimer adopts the icons of exotic religious traditions, quotes from scripture, and cultivates an air of spiritual detachment; but when all is said and done, his claim to fame is a quantum leap in the technology of violence.[50]

The Ritual Cycle of Purification

The atomic bomb, however large it may loom in the histories of the 20th century, is in Wells's book merely instrumental to a far more grandiose vision—a worldwide religion that takes control of human reproduction and eliminates women entirely. This agenda, although seemingly remote from science and technology, is nonetheless completely consistent with the technocratic body image. As shown in the analysis of *Star Wars*, the technocratic symbolic body is a so-called male body, conceptualized as a pair of heads, the phallic and cephalic, which are "created" through a series of cuts that eliminate so-called terrestrial elements and leave behind bloody stumps equated with the "female." Moreover, these so-called male body parts are assimilated to a right/left dualism, represented in the myths as a battle between a pair of twins, which results in the purification of the good twin and the mutilation of his nemesis.[51] Science shares this imagery of pure and polluted, male vs. female, but its special role is to transform the polluting substances into the imagery of celestial light through the creation of powerful ritual objects, such as penicillin and atomic bombs, which simulate the purified body. Thus, to the technocratic player, progress is the transformation of human bodies into the imagery of radiant energy—so human reproduction above all else needs to be put on a higher plane.

In *The World Set Free*, the book which Leo Szilard credits with inspiring his work on the atomic bomb, nuclear weapons are referred to as the "flames no water can quench," and the effects of atomic war are presented, through the eyes of Barnet and the nameless woman, as *death by water*, not death by burning as one would expect.[52] Thus, it is water, particularly "water" from broken dikes and the lower half of the male body (the dead general), and "salt water," such as blood, sweat, and tears, that must be transformed by man-made suns. From a symbolic point of view, Wells's hypothetical transuranium element (93, Carolinum) is a prosthetic device that replaces terrestrial water with its man-made analog; and significantly, the real-life element 93, discovered almost thirty years later, was named *neptunium*—supposedly after the planet but actually for the sea god.

Szilard himself apparently understood that the logical culmination of the scientific process is the replacement of water by celestial fire, for he left Los Alamos after the war to do research in molecular biology; and in this, he was simply following the timetable set down by H. G. Wells. Figure 6.1 gives the dates of the major fictional events in *The World Set Free* in relation to the dates of similar events in the history of the 20th century. In technocratic circles, these correspondences are explained by Wells's historical vision, and one anthology of his work is subtitled "the prophetic books."[53] But do we really have to believe that a minor writer of modest intellectual gifts is one of history's greatest clairvoyants? Symbolic anthropology provides an alternative interpretation. Although the "dates" in H. G. Wells's fictional works may appear to be calendar dates in the 20th century, a perception reinforced by his once widely read nonfiction book *What Happened in History*, they are in fact symbolic numbers mapped to the arrow of historical time. To the anthropologist, Wells's chronology (Figure 6.1) is a carefully constructed liturgical calendar.

As one would expect in a science writer, numbers play an important role in Wells's literary symbolism. Readers of *The World Set Free* are

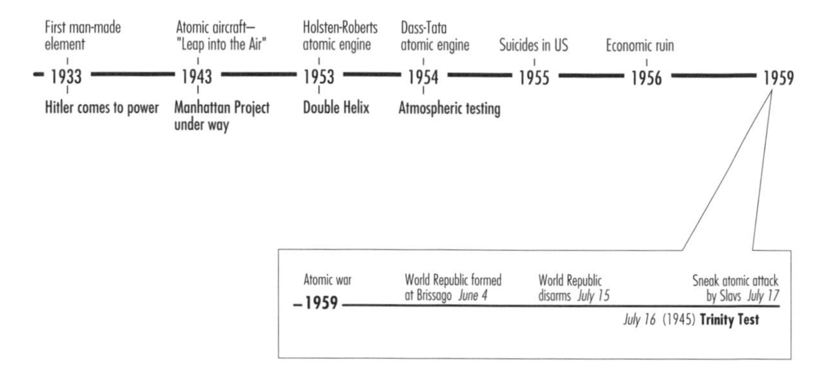

Figure 6.1 Historical and fictional chronologies of 20th-century science. Drawn from information in H. G. Wells, *The World Set Free.*

not only given dates of future historical developments, in some cases down to the month, day, and hour, but a great variety of other numbers as well: the half-life of Carolinum (17 days), the number of dignitaries attending the Brissago conference (93), the years of Barnet's life covered by his memoirs (19 to 23), the number of atomic airplanes produced in France in 1943 (30,000), the cost of running a Dass-Tata engine (one cent per 37 miles), the change in the suicide rate in the United States in 1955 (it quadruples), the number of cities burning in the spring of 1959 (more than 200), and several others. The "historical" dates in Wells's work need to be interpreted in the context of these other numbers, which are symbolic numbers—what I call *numerons*—that perform structural roles in the mythical system. Although Wells's dates appear to denote specific time periods in 20th-century history, they are in fact numerical place-holders in a system of symbolic contrasts.

The majority of these symbolic numbers are not even numbers at all by a mathematician's definition of the term because they are not generated by arithmetical operations, such as addition and subtraction, but are the product of semantic processes, such as the blending of meanings and verbal associations. For example, when the numeron 93 is "divided in half," the answer is not "46.5" as in arithmetic but "9 and 3," as in the two halves of a sandwich. When 93 is "doubled," the answer is not 186 but 9393. The pieces of a numeron can be spatially rearranged like pieces of pottery (9 over 3, 93 versus 39, and so on), but these rearrangements are literal and spatial, and they are not representative of arithmetical operations as they are in normal mathematics. For this reason, most numerons, even though they are numbers with symbolic power, are not equivalent to the use of numbers in numerological systems such as the cabala or physics, for real numerologies use formal algorithms to create and manipulate their symbolic numbers.

The surface structure of Wells's book is a promise of material progress through scientific rationality, so the mythical and ritualistic aspects of his work are necessarily disguised by symbolic transformations of words and images analogous to those described by Freud under the rubric of the dream work. In fact, both the words and numbers in Wells's fictional history are condensed symbols that encode mutilations of the symbolic body, operations which also occur in the process of science itself:

- Fission: split a single, bilaterally symmetrical body part into separate mirror-image parts, e.g., head → left hemisphere and right hemisphere
- Fusion: join together separate but complementary body parts, e.g., eye and eye socket → eye in an eye socket
- Inversion: change a separated, asymmetrical body part into its mirror image, e.g., left arm → right arm
- Duplication: make a copy of a separated body part, e.g., right arm → two right arms
- Swap: exchange two body parts of similar morphology, e.g., egg ↔ roe, or torso ↔ tree trunk
- Transformation: change a body part into the next step in a predefined sequence, e.g., egg → chick

In the symbolic body, both words and numbers can function as body parts, and the language and imagery ultimately determine where the splits are made and how objects are recombined, for the new bodies formed in this way must be meaningful combinations within the system of transformation. Moreover, the fission process imposes certain constraints on recombination, for body parts can only be joined together where there are holes from prior mutilations. For example, if we fission Brissago and Egbert in the following way,

Bris/sago → Bris-, -sago
Eg/bert → Eg-, -bert

then only two new combinations of these pieces are possible, for the "sticky parts" determine where they can be joined together:

-sago + Eg- → Egsago
-bert + Bris- → Brisbert

But the purely logical possibilities of briseg, egbris, berteg, sagobert, and bertsago are not symbolically possible outcomes of the fusion operation because in these cases the "sticky parts" are not in contact. In other words, what I call the *butcher block logic* imposes strict rules on the kinds of chimeras and imaginary words that can be created by repression operations on the symbolic body of a culture, and the one-two punch requires that the prosthesis match the morphology of the amputated body part. For this reason, the analysis of a symbolic body into its iconic deep structure is a rigorously constrained process, analogous to the description of grammatical rules in language.

The numbers in *The World Set Free* are the products of symbolic transformations such as these, and almost all of the "dates" in Wells's chronology can be generated from transformations of just two numerons, 17 and 93, which denote the half-life of Carolinum and the number of delegates at the Brissago conference respectively. These two numbers generate all of the other numbers in the system, including the "dates" of "historical" events, provided that one understands the implicit mythology of death and resurrection conveyed by the narrative. The numeron 93 obviously refers to the outer limits of the periodic table, for in Wells's day there were 92 elements, beginning with hydrogen and culminating in uranium, and the first transuranium element was not created until 1940. So 93 symbolizes the future power of science and the head of Nebuchadnezzar, which is why it is associated with a meeting of heads of state, the good King Egbert, and the formation of a global political system. The numeron 17, in contrast, appears only as the date of the (foiled) atomic attack by the duplicitous king Ferdinand Charles, alias the Slavic fox. Numeron 17 symbolizes the "natural" parts of the male body, that is, the post-cephalic parts that are cut away by successive acts of purification; and this is conveyed by the fact that both Charles and Carolinum have the same name (from *Carolus*, Charles).[54] Together, these two numbers define the extremes of Wells's male body: 93 is white + pure male + head + good, but 17 is dark + natural male + arms + evil.

Wells's chronology (Figure 6.1) is a linear projection of an underlying ritual cycle, which, by its own assumptions, is independent of terrestrial nature and hence unrelated to historical time. From this point of view, the year 1933, for example, the date of the discovery of Carolinum, is not a prefiguration of Hitler's rise to power but a reference to the charter myth of science: 1933 is the 300th anniversary of Galileo's trial, an event which epitomized the conflict between the two major Western religions, science and Christianity. The "year" 1933 begins a sequence of ten-year intervals which mark the development of Christianity's successor: 1933 marks the release of the energy of the atom through the man-made element Carolinum; 1943 begins The Leap into the Air exemplified by the atomic-powered plane; and 1953 is the beginning of industrialization based on the Holsten-Roberts engine, which provided cheap atomic power while changing bismuth into gold. However, from 1953 on, all events are compressed into a single decade, which is characterized by global unemployment, run-away inflation, the ruination of coal mine owners, and tremendous disparities of wealth—social dislocations which lead to war, culminating in atomic war, followed by world government. Thus, 1933 is the first decade in a three-decade sequence, which in turn is part of a larger series of events:

Man-made element → atomic airplanes → atomic power →
world depression → world war → atomic bomb →
Slavic betrayal → world union → world religion →
surgery/death of Marcus Karenin

After 1953, events take place on a yearly basis, with 1957 and 1958 conspicuously missing. Moreover, from 1959 on, the time of events is given only in days, months, and hours, suggesting that the baseline of the chronology shifts as time goes by from "tens of years" to "years" to "parts of years" to "hours of a day."

These "dates" symbolize transformations of the technocratic symbolic body. The two missing years are the time period in which water overwhelms the solar twins (= two suns = two solar years), as indicated by the fact that it is a woman who narrates the story and that Europe is said to be flooded. Conversely, the "date" of the sneak atomic attack "5 A.M. on July 17" is emblematic of a new dawn after which "revolutions" are no longer needed, so solar years come to an end. Moreover, the "date" of "5:30 A.M., July 17" is a complex symbol of the transformed body, consisting of a number of symbolic elements:

- Betrayal (17 = date of foiled attack)
- A regal male head (July = Julius Caesar)
- The buried betrayer (5 = Slavic fox and his 4 companions)
- The winged avatar of the sun ("30,000 planes")
- The Millennium (the 1,000 of "30,000 planes")

These images are not restricted to Wells but are an intrinsic part of the sci-fi tradition which he exemplifies. For example, his symbol of "5 A.M. July 17" could easily be conveyed in the movie *Star Wars* by a scene with the Emperor, the body of Lando Calrissian, and the *Millennium Falcon.*

If Wells's chronology is projected backwards, the next date is 1923, and this numeron in fact occurs in *The World Set Free* as the period of Barnet's life (the 19th to 23rd years) that is described in his memoirs. As shown in Exhibit 6.1, two dates, the extrapolated beginning of the Wells chronology (1923) and its actual beginning (1933), connect the series to the myth of The Lone Galileo. Galileo was tried in the year 1633, and 23 is a numeron in the Galilean myth, one that connects the important episodes in his life. Thus, the hero who is killed and resurrected by the rituals of science is none other than Galileo himself. If Galileo is central to the symbolic body, then one would expect that the numeron 23, along with its arithmetical permutations, would be the most important symbolic number in the entire technocratic mythology, for it connects the microcosm with the macrocosm, interrelating events in the life of the dead hero to the landscape projections created by science.[55] Happily, 23 does everything the anthropologist would want it to do. The terrestrial Galileo (male plus female halves) lives 23 * 2 years from his birth to *Starry Messenger*, whereas the purified body of Galileo (that is, the male half) lives 23 years between *Starry Messenger* and his trial. These numbers also correspond to the diploid and haploid numbers of chromosomes in the human cell, indicating that the life cycle of the

Exhibit 6.1 The life cycle of The Lone Galileo.

Mythic event	Calendar date	Elapsed years between events	Projection
Birth of Galileo	1564		
		2 * 23	Diploid cell (46 chromosomes)
Sidereus published	1610	•	
Trial of Galileo	1633	23	Haploid cell (23 chromosomes)
Death of Galileo	1642	•	•
Body immortalized	•	•	Periodic table (4 * 23 elements)

human cell is conceptualized by genetic biology as a projection of the life cycle of the mythical hero.

However, Wells extends this canonical form of technocratic eschatology by combining it with a literal scenario of death and resurrection modeled on the priest-king of Nemi with allusions to a sacrificed Jewish Messiah, what I have called the myth of Marcus Karenin. In this mythic interpretation of history, Wells transforms the life of Galileo into the death and burial of the priest-king, conceptualized as a pair of twins that symbolize the temporal and spiritual aspects of Galileo respectively. As shown in Exhibit 6.2, the temporal and spiritual halves of Galileo begin their separation in 1633, the year when he was placed under house arrest; and the mortal coil lingers on for nine more years, whereas the spirit goes underground for a 300-year latency period.[56] During this time, the mortal or temporal form is transformed into an immortal, man-made body, namely, the periodic table, whereas the immortal aspect finally reemerges as the spiritual Karenin.

As a new international religion analogous to Christianity requires a self-sacrificing Jew, Karenin must shed his latent Jewishness in order to usher in the new age. In the technocracy there are no metaphors, only physical bodies, so Karenin's transformation takes the form of a literal surgery to remove his congenital deficiencies. The exact nature of the surgery is unspecified in the book, but in technocratic circles, the phallic fork or procreative organ, while profoundly masculine in contrast to women, is seen as a source of pollution when compared to the radiant cephalic fork, which is celestial in origin and independent of nature. So real spirituality requires castration, which is to say, the removal of natural procreative abilities and their replacement by prostheses. In fact, in the technocratic symbolic system, castration, symbolic or otherwise, is equated with spirituality. Moreover, *testament* is from Latin *testari,* "a will," a usage that occurs in English as *intestate,* "to die without a will," which implies a lack of legal heirs, and heirs are normally the fruit of one's *testicles.*[57] Therefore, to a literalist, New Testament is New Testicles. Not surprisingly, the head surgeon is named Fowler (associates to *cock,* Latin *gallus* > *Galileo*). Thus, in Wells, "spirituality" goes beyond the asexual monasticism of *Star Wars* and is equated with a literal self-castration by a self-sacrificing Jew, in which the cephalic fork of the Y-forked organ (white-coated head surgeon in Tibetan monastery) operates on the phallic fork (Red Jewish brain in congenitally deficient body surrounded by women)—an operation which elevates the technocratic system to the ultimate in self-sufficient masculinity.

Exhibit 6.2 The life cycle of Galileo the priest-king.

Mythic event	Calendar date	Elapsed years between events	Projection
Birth of Galileo	1564		
		2 * 23	Diploid cell (46 chromosomes)
Sidereus published	1610	•	
Trial of Galileo	1633	23	Haploid cell (23 chromosomes)
Death of temporal Galileo	1642	•	•
Temporal body immortalized	•	•	4 * 23 (periodic table)
Wells's birth	1866	233	Cell nucleus split (years from 1633)
Spirit head germinates	1933	•	Wells's element 93
Airborne seeds released	1943	10	Wells's Leap into Air
Hands and trunk differentiate	1953	10	Wells's twins, war, etc.
Man-made body formed	1959	2 * 3	4 * 23 + 1 (Periodic Table = Brissago)
Twin heads differentiate	July 17	•	Sneak atomic attack
False head killed	•	•	Slavic fox killed
Congenital disease cured	moonset	•	Karenin's surgery
Blood from wound stops heart (Artificial gonads replace blood and biological rhythms)	•	0.01923 (a week)	Karenin's death

The development of this asexual "spirituality" in Wells is mapped to the growth of the symbolic body, which germinates in "1933" and develops through a series of stages, reaching its mature form sometime after "1959." It is important to remember that the hard-and-fast distinction in the public eye between biology and physics is itself a creation of the ritual process, without significance on a deep structure level, and that technocratic players such as H.G. Wells and Warren Weaver saw atomic nuclei and cellular nuclei as symbolic equivalents. An "Atom" is an "Adam" in this system, because both are man-made celestial fires that water cannot quench, and both can be split by science to create a pure "half-life," namely, radioactivity in physics and the haploid cell in biology. So the release of energy from the "atomic nucleus" envisioned by Wells was just as much an evocation of the past as a prediction of the future, for the cell nucleus had already been split, about the year that Wells was born, by the chemist Friedrich Miescher.[58]

In modern biology, the significant genetic information is considered to be in the cell *nucleus,* a hard, cephalic element at the center (derived from Latin *nucella,* "little nut") which imposes form on the peripheral cell body or cytoplasm (*cyto* from Greek *kytos,* "a hollow place"). In the origin myths of canonical science, nucleic acid (RNA and DNA) was discovered, not by Watson and Crick, but almost a century earlier, when Miescher isolated it from the nuclei of white blood cells, which he obtained from pus from infected wounds. White blood cells are not only white, as opposed to red, but they do battle with alien microbes—AIDS is due to an insufficiency of them. Nucleic acid is thus associated in its own scientific origin myth with the "white" "nuts" that defend the body against alien invaders. This origin myth of DNA is in fact replicated in Wells's story, for it is the French diplomat Leblanc ("the white") who organizes the conference in the Alps, that is to say, builds the new body; and it is he who defends the gathering against the duplicity of the Slavic fox by secretly "doubling its air defenses" (that is, by making a pair of purified arms where one was before—a pair of duplicities).

In the technocratic symbolic body, the active genetic principle, although it lies dormant in nuts, also moves through the air as a floating sphere with radiant symmetry, like dandelion (="lion's tooth") down or a parachute. It was a floating seed, after all, that supposedly came in Fleming's window and killed the microbes in the petri dish. Thus, the first stage in the seeding process is The Leap into the Air, an image captured perfectly in Woody Allen's movie *Everything You Always Wanted to Know about Sex (But Were Afraid to Ask)*, in which sperm cells are portrayed as parachutists leaping into the breach. In Wells' chronology, The Great Leap into the Air (1943), coincident with the 400th anniversary of Copernicus's heliocentric theory, is a period of forward movement in which atomic power is used to propel airplanes, of which 30,000 were produced in France alone. But this progressive period is soon followed by an atomic bomb blast over Paris, which leads to breached dams, and a flood which inundates all of Europe. Symbolically, this an elaboration of the semantic associations implicit in the Holsten-Roberts engine, as shown by the following butcher block logic:

Holsten-Roberts engine ⟶ Hols-
-ten-
-berts
Ro-
eng-
-ine

b/e/r/t/s ⟶ terb + s

terb + ine ⟶ terbine

e/n/g ⟶ gen

yielding, in conventional spelling:

ten holes + roe + turbine + gen[erator] + S

Thus, in technocratic imagery, "nuclear energy" of both sorts is associated with leaping aquatic creatures, the transition from salt water to fresh water, with fish eggs, turbines, and breached dams. The ever-dependable Miescher performs on cue: after isolating nucleic acid from white blood cells, he confirmed his initial research with the sperm of spawning salmon[59]—from Latin *salmo,* "the leaping fish," cognate with Latin *saltare,*"to leap."

The nuclei of leukocytes, Miescher's white nuts, are important to H.G. Wells as well, for in another of his sci-fi books, *Men Like Gods* (1923), the absence of an immune system is taken as a measure of technological progress: "for more than twenty centuries the Utopians had had completest freedom from infectious and contagious diseases of all sorts... And there followed a corresponding change in the Utopian physiology. Secretions and reactions that had given the body resisting power to infection had diminished: the energy that produced them had been withdrawn to more serviceable applications."[60] Wells's position, however bizarre it may seem to a normal disease-prone inhabitant of the planet earth, is perfectly consistent with the technocratic agenda of biological perfection, for it is technocratic structures that protect the body from disease, not the body itself. So the absence of an immune system is another way of saying that the body has been transcended and replaced by celestial light.[61] Moreover, the very concept of an immune system is threatening to the Y-forked organ, for the leukocyte, while admittedly a *white* blood cell, is still a *blood* cell, whose highest creation is pus—hardly the sort of golden head that could be expected to vanquish enemy aliens. In the symbolic body of the technocracy, white blood cells, like sex itself, demand to be replaced by man-made organs.

Who Killed Laurel/Palms?[62]

So physics creates the purified hands with which genetics constructs the purified body. In the androcentric mythology, arms and hands are "male" organs, as in military *arms* and right-*hand* man, ranking just below the head in social status. The hands constructed by physics are the man-made actinide elements from 94 on, which provide the radioactive isotopes used to trace biochemical pathways and the X-rays used to image macromolecules. Significantly, Wells's new age begins with the creation of element 93, a transuranium element; and in conventional histories of science as well, the "atomic age" is ushered in with the fissioning of uranium and the discovery of plutonium. As the chemical elements 1 to 92 occur naturally on the planet earth, they are less pure than man-made elements, which occur naturally in stars, so in charts of the periodic table the transuranium elements, along with thorium, protactinium, and uranium, are spatially separated from the body of the chart, forming a separate appendage, the actinide series. Thus, the transuranium actinide elements, with the exception of neptunium, are the arms and hands of the new and purified body, which protect it against external enemies and enable it to construct the prostheses that will replace the terrestrial prototypes. Moreover, the actinides are divided, like everything else in the system, into two mirror-image halves, dexter vs. sinister; and the sinister hand, being closer to "female" "nature," needs to be purified by fire. Also, digits, as we know from anatomy, are really a formation of Persian infantry (phalanges) and thus clearly male, so the symbolic system classifies palms and digits as logical opposites.

The somatic contrast between the noble male head crowned with laurel and the less noble palm rooted in the earth is conveyed by the names of the transuranium elements, almost all of which were named by one person, Glenn T. Seaborg: discoverer of plutonium, alumnus of the Manhattan Project, professor of chemistry at Berkeley, Nobel laureate, and former chairman of the United States Atomic Energy Commission. With a pedigree like this, it is hardly surprising that his nomenclature of transuranium elements is a textbook case of laurel vs. palmar. Figure 6.2 shows how the right "male" hand of the transuranium elements, which is higher up the scale (99–103) than the left "female" hand (94–99), is composed entirely of male heads: Nobel laureates (einsteinium, fermium, lawrencium), Nobel himself (nobelium), and the human creator of the prosthetic body (mendelevium).[63] The left hand, in contrast, has no solitary male heads at all but is a diverse mix of plutonium, americium, curium, berkelium, and californium. Three of these names are places on *earth,* namely, America, California, and Berkeley; one is under the earth (plutonium); and the fifth is a male/female pair named for Pierre and Marie Curie. As every entity in the technocratic system must be higher or lower than something else, the right hand is conceptualized as transforming the digits of the left hand by fire, leaving a purified palm, with radiant energy replacing the digits. Significantly, the real-life Madame Curie, née Sklodowska, also had hands purified by science, for as the histories of the A-bomb never fail to tell us, they were blotched and withered from handling too much radium.[64]

Simply by assuming that the digits of each hand form a series, it is possible to put the fingers of the two hands into a one-to-one correspondence with the transuranium elements, such that each finger points to its symbolic opposite. No matter how the two hands are oriented, the middle finger ("fuck finger" in idiomatic English) of the right hand points to the middle finger of the left hand. That is, the artificial body (mendelevium = Mendeleev = creator of the periodic table) penetrates the natural body (curium = Pierre and Marie Curie = a peter and a mary = scientific prodigy).[65] However, there are four possible orderings of the other fingers (left thumb = Cf *or* Pu *and* right thumb = Lr *or* Es), leading to four different symbolic interpretations. Of these, the ordering that is most consistent with other aspects of the system is thumbs = Lr and Pu. In this interpretation, Einstein (Jewish brain and pinky) transforms plutonium's bad smell (that is, tainted money); Fermi(um) both connects the Old and New Worlds and transforms the latter (americium) into a world power; Nobel-ium prizes make Berkel(ey)ium into the center of the universe; and Lawrence(ium) transforms Californi(um) into the promised land.[66] Whatever the exact correspondences, the resulting image, as shown in Figure 6.2, is a mimicry of Michelangelo's Creation of Adam on the ceiling of the Sistine Chapel; but in this new, scientific vision of creation, God is replaced by Nobel laureates, the earth by pollution, the human body by disembodied heads and hands, and the spark of divine life by nuclear explosions.

Now, forty years later, this technocratic act of creation is to be made literal by constructing a new biological body using the gloved, irradiated hands of the purified female. Crick said as much in 1957: "The two chains of DNA, which fit together as a hand fits into a glove, are separated in some way and the hand then acts as a mold for the for-

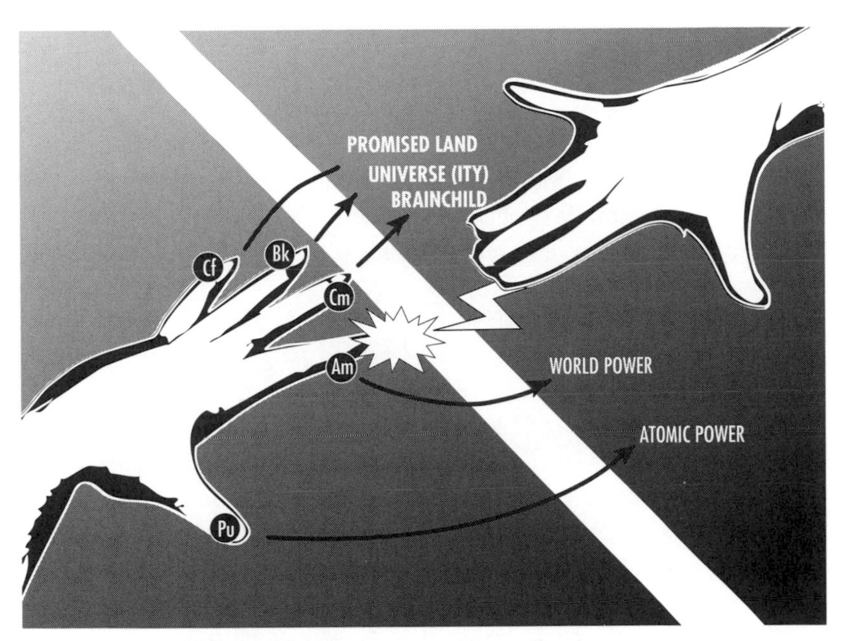

Figure 6.2 The transuranium elements are the hands of the technocratic symbolic body.

mation of a new glove while the glove acts as a mold for a new hand. Thus we finish up with two gloved hands where we had only one before."[67] *Mold,* of course, has a double meaning in English—*mold*[1] denoting a hollow shape that imposes its form upon a plastic substance with which it is filled, and *mold*[2], a downy or furry growth on organic matter or the fungi which cause such growths. In this condensed image, the hand is the mold in sense 1, and the glove is the mold in sense 2. The development of the human body in this androcentric symbolic system is exactly the opposite of what one would expect on the basis of familiarity with normal terrestrial existence. In normal life, hands and body are biologically connected to each other, but in Watson's crick model, the "hands" are nuclear material, transferred from one generation to the next without contamination by the body, which is conceptualized as peripheral and moldy. These self-contained "male" organs, the "hands," build the rest of the body from the center out, assembling proteins through a complex series of cut-and-paste operations controlled by instructions from the center. Given this set of assumptions about the human body and its relationship to nature, it is hardly surprising that the technical processes of contemporary genetic engineering replicate the Manhattan Project:

> First of all, *fission a cell nucleus* to remove the *nuclear material.* Then "explode" it into its constituent pieces. Next create a "chain reaction" by producing multiple generations from each DNA fragment (cloning). Attach *radioactive isotopes* to the cloned fragments, then reinsert them into normal DNA, producing a man-made gene whose activities can be monitored by radioactive exposure of photographic film.[68]

In His Image

The late 1970's was the time of transition between science fiction and science fact in the field of biotechnology; and as with its predecessor atomic energy, the vision of the future was first articulated in sci-fi novels and popularizations, without ever appearing in textbooks in its overt form. Wilson's *On Human Nature* highlighted immortality as the one area of traditional religion that remained to be conquered by science; and that same year (1978) a science journalist, David M. Rorvik, published *In His Image: The Cloning of a Man*, a work that intentionally muddles the boundary between fact and fiction.[69] Rorvik claimed that his book is a work of journalism, even though there are no specific names and dates in the narrative, and the title page carries a disclaimer from the publisher. But whether journalism or science fiction, Rorvik's work is a window into the contemporary technocratic agenda.[70]

The book tells the story of a childless, rich industrialist, denoted only as Max, who is both an orphan and a twin. He successfully clones a male heir using the techniques of molecular biology and the artificial implantation of a cloned fetus in a woman's womb. Max's hero is Hermann Muller, the Nobel laureate in genetics who advocated the creation of human gene banks, artificial insemination, the economic exchange of ova, and the freezing of embryos for later use—in order to circumvent political objections to his vision of state-controlled breeding of a genetically engineered master race.[71] In the story, Max hires Rorvik, the author of the book, to find a scientist with the skills to perform the first cloning of a human being. Rorvik locates a man code-named Darwin who is willing to undertake the secret cloning project. Where Max is a near-vegetarian and a jogger,[72] Darwin tends to be overweight,[73] and his hero is Henry Kissinger, recipient of the Nobel Peace Prize and the architect of the secret bombing of Cambodia.[74] (Thus, the characters' heroes are Little Boy and Fat Man, the Nazi and the Jew.) In 1974, Max builds Darwin a state-of-the-art biotech laboratory in an unnamed Third World country. The laboratory is disguised as a hospital for local women, whose wombs are used for the experiments under the guise of medical treatment. Darwin is assisted in this effort by a female gynecologist named Mary. A stable of healthy virgins is assembled as potential surrogate mothers, and after some initial failures, the man-made fetus is successfully implanted in the woman code-named Sparrow, a perfect virgin except for a burned right hand. Max insists that the baby be born in the United States, so Sparrow, concurrently with her pregnancy, is educated in "American culture."[75] The pregnant woman is then flown to California in December 1976, where she gives birth to a healthy male heir. She is said to be disappointed that the baby is not born on Christmas day, rather two weeks before; but Max is pleased, for he sees the child as his contribution to the bicentennial year of the United States.[76] Whether Rorvik's story is science fiction or science fact, it references all of the themes of the Manhattan Project—such as, the twins, the secrecy, and the man-making process—but the imagery of the Oppie Effect has been escalated to a literal virgin birth.

Another paradigm work in this emerging tradition is Brian Stableford's *Future Man,* published in 1984, which presents biologized eschatology in terms of vivid, full-page illustrations of a genetically altered future.[77] In one illustration, a large blue pool, reminiscent of the cooling tank in a nuclear reactor, is filled with maturing human fetuses, each in its own glass bubble, and the caption promises a new (and final) stage of human evolution in which genotypes are the result of choice, not chance.[78] Other pictures show hypothetical human phenotypes that have been genetically engineered for submarine and extraterrestrial habitats, as well as genetic warriors bred for war.[79] The chapter titles alone indicate the broad scope of the program: Mastering Our Environment, Spare Parts for People, Control of the Mind, Extensions of Man [prosthetic organs], and so on.

In the symbolic body encoded in the atomic bomb, physicists are at the center looking out (gleaming spheres), whereas biologists are on the outside looking in (ugly brown bugs' eyes), but now the ecliptic of the Double Helix is occluding the globes of the Double Sun. Given the intrinsic relationship between the scientific world view and the symbolism of purity and pollution, we would expect the medical profession to play a central role in the implementation of technocratic agendas, just as it did in Nazi Germany; and not surprisingly, many features of the new eschatology have already emerged in the United States—presented, of course, as "neutral techniques" in the realm of medicine, such as *in vitro* fertilization, synthetic organisms, the commercial use of "surrogate mothering," cryogenics as a technique for achieving immortality, and, most important of all, the Human Genome Initiative, which in the words of its own promoters is to biology as the Manhattan Project was to physics.

The Human Genome Initiative is a massive undertaking, international in scope, involving the funding of research in dozens of countries and a computerized database at the Los Alamos National Laboratory. It has as its objective the sequencing of all of the genes in the human genome, that is, a computerized description of all the base pairs in the DNA molecules of the human species that are known to encode genetically transmitted information. This research, with an estimated price tag of three billion dollars, is being sponsored by the United States Department of Energy (DOE), formerly the Atomic Energy Commission, the government agency charged with the development and manufacture of nuclear weapons.

The project is being funded, we are told, to cure genetic diseases. In a comprehensive article written in 1988 for the respected journal *American Scientist*,[80] the Chairman of the Department of Biomathematical Sciences at the Mount Sinai School of Medicine in New York, for example, lists the benefits, both medical and technological, that will follow from this research. The author is surprisingly candid about the role that Big Biology is to play in future technological progress, claiming that the DOE workshop on the Human Genome Project generated an enthusiasm among the participants that was reminiscent of the Manhattan Project.[81] But no problems are anticipated, for, as the author assures us, the U. S. Department of Energy has had responsibility for studying the health effects of radiation since the agency was founded at the end of World War II.[82]

But wait—is not this the same agency whose great regard for human health led them to cover up for more than twenty years the irradiation of the towns in Utah by fallout from Yucca Flat? The agency that supervised the secret venting of radioactive iodide over eastern Washington State? The government body which prevented the Public Health Service from warning the American public about the dangers of radiation? And have not all these facts been documented by declassified government documents and sworn testimony in American courts of law?[83] But now, this public servant, along with Los Alamos National Laboratory, the venue of the Manhattan Project, and Lawrence Livermore Laboratory, founded by Edward Teller to design his little Mikey, although not exactly hospitals, is engaged in a multibillion-dollar program to improve the genetic well-being of the human species, irrespective of color, creed, or national origin.

As the postwar symbolic body shifts, moving the imagery of A-bombs to the periphery and bug eyes to the center, the relative status of physics and biology shifts too, giving "life science" the place once held by physical science, with the mandate to create the society-wide images that dramatize Nebuchadnezzar's Dream and to build the artifacts that protect us from enemy aliens.[84] But as we have seen, molecular biology is in a complementary structural relationship to physics, mechanizing life while physics vitalizes matter, so if the androcentric cult of physics is expressed by the secret construction of Y-forked organs, then the imagery of molecular biology is expressed by the secret construction of *complements* to Y-forked organs. In the androcentric cult, however, the complement of a male organ is not a woman—God forbid!—but rather, on analogy with plutonium, a man-made "female" substance (P.U.!) that is then transformed by means of lenses into a prosthetic "female" organ. Thus, biology must produce a "soft bomb" to complement the hard, cylindrical shape of an intercontinental ballistic missile with its nuclear warheads. Yet even critics of biotechnology have been so diverted by the wind-up hare of "ethical issues" and "social responsibility" that no one apparently has ever even asked the obvious question as to why laboratories of nuclear weapons development would want a database of genetic markers that uniquely identifies every human population on earth.[85]

It appears that Hollywood scriptwriters have a far deeper understanding of science than science historians and science journalists, for the one-two punch of physics and biology has long been recognized in the movies as an ominous development, whereas the serious press and the academic establishment think that the objective of genetic research is a medical breakthrough that will cure disease and end the pain of childbirth. But from a symbolic point of view, HGI is the eye of Mercury (Hermes)—the master of disguises who curses the sun in *The World Set Free*—and its function is to build the dark invader Anakin from the mutilated remains of the dead king. That is, the hermetic Mercury (Hg) connects the mythology of *The World Set Free* to its modern manifestation *Star Wars*, in which the sacrificed priest-king is conflated with the Messiah of Judeo-Christianity, resurrected from the dead by means of the genetic control of reproduction, and sold to the public as Lucas/ Spielberg stews of doves, fishtails, and Lost Arks (Exhibit 6.3).

Exhibit 6.3 Lucas's *Star Wars* and Wells's *The World Set Free* exemplify the same iconic deep structure.

The name *Marcus Karenin* is an anagram of *Anakin* plus *Mercury* in which the *Y* of *Mercury* is replaced by an *S*.

M/a/r/c/u/s K/a/r/e/n/in → Anakin + Mercur (y → s)

The latter substitution is symbolically appropriate because the god Mercury is conventionally depicted carrying a staff around which a serpent is twined in an S-shaped configuration. That is to say, Mercury is the god whose Y-forked organ is replaced by a serpent-rod.

Crisis of Faith

In *The Handmaid's Tale,* the novelist Margaret Atwood creates an imaginary future in which Fundamentalist Christians join forces with paramilitary groups to form a fascist-like dictatorship in which women are forced to be breeders and housewives.[86] Few American reviewers found this story credible, nor did the movie do well at the box office, probably because Atwood's concept of sexual oppression assumes that women's bodies are the source of life—an idea which the technocracy long ago transcended.[87] In the scientific mythology, the source of life is celestial light transformed by a phallic apparatus,[88] and it demands the *replacement* of women's bodies by disembodied structures rationally conceived.

Atwood's book is in the liberal tradition, which sees religion in general and Christianity in particular as the enemy of the secular state and of civil liberties; and liberal social theory would view the concept of a scientific eschatology as a contradiction in terms. Yet it is science that fosters Nebuchadnezzar's Dream; and it is science the established religion, not Christianity, that has the most to lose from a free market of ideas—so it is science and related institutions that are most likely to use the law to enforce their world view and preserve their privileges. In this bid for intellectual monopoly, the technocracy is aided by the postwar secular state itself, for, far from being a neutral structure with a hands-off attitude to theological questions, it has become a covert partisan in an essentially theological debate. Science, as we have seen, is not "a secular alternative to religion" but a real religion that promises immortality to its practitioners and covertly references theological propositions. The secular state, instead of fostering a climate where no one religion can dominate the state apparatus, now uses its power and prestige to reinforce the technocratic world view by explicitly excluding all religions but science from public life and by actively fostering the scientific cosmogony through its support of basic research, technological development, and mandatory scientific education. Hence, Atwood has discerned the latent theocratic theme in contemporary American culture, but her liberal premises force her to interpret it within a 19th-century framework as a Christian usurpation of the secular state, manifested by direct political control of real live wombs. Consequently, she fails to see that the logical development of the covert relationship between the modern state and scientific eschatology is not a *Christian* theocracy but a *technocratic* one.

As many feminist writers themselves have pointed out, the political control of female reproduction is highly compatible with technocratic ideology, for it conforms to the underlying imagery of bringing "female" nature under rational control. Davis-Floyd, for example, challenges what she calls the "technocratic model of birth," which she interprets as ritual acts that function more to display the technocracy's image of women than to facilitate the health of the mother; and she shows through interviews with pregnant women that the androcentric concept of birth as a disease is accepted as right and natural by many American women themselves. Gena Corea, in an extensive review of contemporary biomedicine, sees the new technologies of reproduction as a conscious attempt to remove women from the birth process entirely.[89] This same theme is latent in many other books written from a feminist perspective, and even Atwood's "the handmaid's tale" can be read as a covert reference to "hand-made tails."

Anthropological analysis not only confirms these suspicions but it indicates that the contemporary technocracy is far more insidious than most feminist critics have recognized. Feminism presupposes the existence of two sexes, male and female; and locker-room macho behavior, for all of its vociferous assertion of male uniqueness, still assumes the existence of *women*. But the technocracy, even though it is androcentric in its imagery and patriarchal in its distribution of power, is antithetical to *both* sexes. Biological reproduction, with its messy blood and mucous, is seen as incompatible with what a real human being should be; and the final stage of human evolution is construed as a bloodless, fearless, and disembodied state in which "nature" is transformed into radiant energy. In *Star Wars*, biological bodies, both male and female, are viewed as sources of contamination and pollution, and the most "human" characters in the film in terms of their motivations and feelings are quasi-disembodied spirits like Ben Kenobi or full-fledged machines like C-3PO and R2-D2. Conversely, "spirituality" is portrayed as disembodied mental capacities, such as psychokinesis, telepathy, and clairvoyance, in which spaceships are raised from the swamp by the power of the mind alone. This interpretation is confirmed by the cults of technocratic eschatology developing now in the United States, described by Regis in *Great Mambo Chicken and the Transhuman Condition*.[90] Their premise is that the individual human person is no more than patterns of information in the brain, for which the biological body is a flawed and superfluous chassis. Thus, they promise the resurrection of frozen brains by robotic cryogenicists, the down-loading of minds to computer programs, and the beaming of backup copies of the self to the furthest regions of outer space.[91] In short, "androcentric" imagery is no longer "male" but the vehicle of disembodied androgyny.

The trend towards technocratic eschatology, initially expressed as immortal intellects floating free in outer space, is already well developed in the United States, prompted by widespread disaffection with the materialist vision of traditional science.[92] In earlier periods of history, from the 17th to the mid-19th centuries, science was more of a promise than an achievement; and any shortfalls in the earthly paradise could be attributed to an insufficiency of the technocratic spirit. Until the 1830's, with the exception of a few inventions, such as the Newcomen engine, improved pumps, better guns, and the marine chronometer, most scientific technology was confined to the laboratory; and the technological spin-offs—that is, the ritual objects that display the imagery of the myths—did not significantly affect the lives of ordinary people.[93] In the 19th century, however, the material world of the average person in Western Europe began to change under the impetus of industrial research and development: electric lights, cheap steel, railroads, internal combustion engines, synthetic fibers, aniline dyes, TNT, and a host of other innovations, all made possible by science and engineering. At the present time, science defines the world in which we live, not just in material goods, but, more importantly, in the kinds of questions that can be asked and the kinds of answers that are acceptable. Every major political issue requires a scientific poll to tell us what we think, and every box of breakfast cereal boasts statistics on health and nutrition. But the more that the technocracy succeeds in material terms, and the more that its rationalist and mechanist premises become the dominant metaphor of human life, the more obvious the discrepancy between its eschatological claims and its ability to deliver.

Modern science, after all, does not justify itself in terms of marginal improvements in the health and comfort of ordinary citizens—to the contrary, even to its own ideologists, such technological benefits as toasters and guided missiles are firmly in the realm of engineering and *applied* science, quite distinct from real science, which is *pure.* Rather, the materialist engine is fueled by an implicit atheological agenda that appropriates Judeo-Christian symbols of resurrection and deliverance and recasts them in physicalist terms. By short-circuiting spiritual, ethical, and metaphysical questions, the technocracy promises to make a new, rational, well-ordered world where evil, death, and disease are physically impossible, a kind of Federal Express of the soul, delivering overnight what with religion took an eternity. But it is one thing to offer people running water and electricity, quite another to provide them with happiness and immortality.[94] The former claims are firmly in the realm of craftsmanship and engineering, amenable to empirical methods, but the latter are intrinsically metaphysical constructs, firmly in the domain of traditional religion. Thus, the more successful science is in altering the material conditions of life, the more its spiritual limitations become apparent. Already the erosion of the scientific vision is being felt by the mechanistic faithful who are desperately waiting for a technological breakthrough that will snatch them from the jaws of death (Exhibit 6.4).

As scientific premises are increasingly challenged on both a philosophical and pragmatic level, and as the technocracy is under increasing public pressure to deliver its long-promised immortality, then traditional science is put in an impossible position. Either traditional science admits that it is just an opportunistic process of technological development devoid of spiritual content, or it publicly affirms its latent eschatological agenda, losing in the process the protective cloak of secular knowledge, which gives it the privilege of debunking theology in the public schools. Either way, the spiritual credibility of science will be shattered and its political usefulness at an end.

Tumor victim sues to have his head preserved. He hopes freezing will give science time to find a cure.

T— D—, 46, of Sunnyvale files his lawsuit Monday in Santa Barbara Superior Court. It says he wants his head "cryonically suspended" in hopes that scientists will discover a way to attach it to a healthy body and cure his brain disorder. Cryonic suspension is an experimental procedure in which all or part of a person's body is preserved at minus-320 degrees Fahrenheit. To date, the procedure has been performed only after a cryonics patient has died of natural causes. D— wants to die before the tumor significantly damages his brain. His request, if carried out, would amount to suicide.... If he wins his suit, D— would be anesthetized and placed on a heart-lung machine, which would pump blood through his body while his temperature was reduced. His blood would be replaced with chemicals to protect his cells from freezer damage. D—'s head would be surgically removed at the end of the operation. His head would then be placed in a Thermos-like device and maintained at a temperature of minus-320 degrees Fahrenheit....

Note the equation of the brain with the body and the person with the brain.[95]

Exhibit 6.4 An example of technocratic eschatology.

But as science ceases to be a vital intellectual force that can challenge religion on its own ground, then the functions of science will have to be assumed by the technocracy itself, for the whole international order has been built on the linear arrow of progress, with its implicit promise of materialist salvation. As symbolic analysis shows, scientific theory and technocratic mythology are not identical. The technocrat is a player not a mechanist, with a far greater commitment to the atheological agenda and the imagery of progress than to the facts and theories of basic research; and Galileo himself is valued more for challenging the authority of the Church than for his contributions to physics. Moreover, scientists are fast becoming dispensable. The long-hairs and eccentrics, the pinkos and propeller-beanie boys, are far less important than they once were, for organized religion appears to have been intellectually discredited, and the technocracy now has effective control almost everywhere. Science may have been essential in the creation of the modern world, but it is not essential to the maintenance of it. With the exception of economic rationality, bourgeois institutions such as science are more liabilities than assets to contemporary ruling elites, for they engender endless irrational objections to the clean, well-lighted place envisioned in the plans. Moreover, the scruples of old-fashioned scientists are in many respects a serious obstacle to the next generation of technocratic goals. However limiting its intellectual premises, the scientific method encourages attention to facts, stresses the importance of evidence, and inspires a concern for truth and falsity—habits of mind which can only inhibit full-blown technocratic fantasies. Surely the technocrats remember that it was scientists who first attacked the Strategic Defense Initiative ("Star Wars") and testified to Congress that it probably would not work. Moreover, the technocracy thinks that it already has the Secret of Life, one so simple and straightforward that it can be down-loaded to technicians and managed by MBA's.[96]

More importantly, the underlying logic of technocracy requires it to recreate religious iconography in physicalist terms. Whereas the committed materialist is content to live in a stripped-down world of scientific facts, the player cannot rest until Judeo-Christian imagery has been reproduced by rituals of reduction. This can be done covertly in the case of cosmogony, but the reduction of eschatology is far harder to reconcile with the theories of conventional science, for what to the materialist are the most absurd aspects of traditional religion, say the transformation of wine into blood or the resurrection of Christ from the dead, is to the player the ultimate technical challenge. For if science is the new religion and the heir to Judeo-Christian mythology, then it too requires a Messiah in human form who speaks with the voice of God, is sacrificed, and rises from the dead. But as the technocracy is a physicalist parody of religion that can provide no real spirituality or formulate any ethical standards, it opts for the imagery of theology while ignoring the moral and philosophical content. In Christianity, the Father is the source of justice, the Son the source of redemption, and the Holy Spirit the bestower of the cardinal virtues of faith, hope, and charity; but in the technocratic parody, the three persons of God are expressed by purely iconic properties that lend themselves to the abuse of power:[97] The Son, a man-made messiah offering light as the source of eternal life; the Father, the male voice from the sky with the power of life and death; and the Holy Spirit, a dove that speaks all languages. In *The World Set Free*, after all, atomic war is followed by a world religion conceptualized as a successor to Christianity.

The imagery of technocracy indicates that a logically consistent eschatology, couched in the terminology of biological science, is currently building in the subculture of laboratories, medical institutions, and government agencies, both in the United States and in other countries with a heavy commitment to the imagery of technocracy. Although the system is not yet institutionalized, the surface structure imagery of the popularizations indicates a coherent constellation of beliefs that can be summarized as follows:

- Human nature must be superseded if the species is to advance, so we need to take control of the evolutionary process by means of molecular biology.
- However, sexual coupling between men and women for reproductive purposes is a primitive technique that perpetuates sexist relationships.
- To be liberated from sexism, women must abandon childbirth in favor of asexual reproduction based on more modern, scientific procedures.
- Moreover, as speciation occurs in times of environmental catastrophe, and as nuclear war induces genetic change, then the new age will result from a holocaust—for the selected are the elected.

This theology offers a purely physicalist concept of spirituality that obviates the need for any serious changes in our concept of human nature or in the relationship of the human species to the earth. Since human nature is defined in purely instrumental terms, as the control of biological evolution, the theology reinstates the unbridgeable boundary between nature and culture. It also reinstates the disembodied head—the control of biological processes by abstract intellectual structures. Moreover, by replacing conventional religions with one of its own, the technocracy can proffer a concept of spirituality unencumbered by morality, ensuring that social relations can continue to be ignored, while giving its own landscape projection free reign. Also, the mythical reformulation ingeniously perpetuates the imagery of women as leaky vessels—by offering to mend them, and love between the sexes is even more irrelevant to the new technocratic agenda than it was in the old. Last but not least, the new spirituality manages to put a positive spin on nuclear war, seeing it as the evolutionary agent of transformation that will catapult the human species to the next stage of cosmic consciousness.

However, as society is still officially secular, the technocratic parodies of religion continue to adhere to the taboo of a strict separation between the scientific treatise and the popular polemic. Even flamboyant rituals of biological assault present themselves as practical medical techniques, while manifestly technocratic eschatologies disguise themselves as sects of Christianity. A good example of the latter is the cult of the Rapture. This is an eschatological theme within some Christian Fundamentalist churches, in which the true believers in Jesus, those who are Born Again, will survive nuclear war by being transported bodily into heaven as the bombs go off, leaving unbelievers to be consumed in the conflagration.[98] Although it presents itself as a form of Christianity, with references to the Book of Revelation, the Rapture, like so much of contemporary Fundamentalist Protestantism, has far more in common with the imagery of radiant energy than with the mandate of loving one's neighbor. Not only does the cult have strong ties to communities that manufacture nuclear weapons, but a man-made, technological event is conflated with the symbolism of the second coming of Christ. Could there be a more unambiguous example of the Oppie Effect than the equation of nuclear weapons with the second person of the Christian Trinity?

However, as overt eschatology is completely incompatible with science, which must deny even the possibility of a theological perspective, the more scrupulous scientists will be pensioned off; and the underlying mythology of industrialism, under strong public pressure to deliver its long-awaited immortality will mutate from a covert to an overt religion, unconstrained by mechanism on the one hand or by popular support for traditional creeds on the other. E. O. Wilson, professor of biology at Harvard University, expresses the theocratic agenda of genetics in words that could have been uttered by Marcus Karenin himself: "Like the mythical giant Antaeus who drew energy from his mother, the earth, religion cannot be defeated by those who merely cast it down. The spiritual weakness of scientific naturalism is due to the fact that it has no such primal source of power. While explaining the biological sources of religious strength, it is unable in its present form to draw on them, because the evolutionary epic denies immortality to the individual...." "Does a way exist ," he asks rhetorically, "to divert the power of religion into the great new enterprise [genetic biology] that lays bare the sources of that power?"[99]

Exercise 6.1 Construct a biological analog to the Manhattan Project.

Use bioengineering techniques, the iconic pathway of World War II technology (Figure 5.3), and biological examples of the following symbolic propositions:

- Man-made white blood cells (Lucas cytes) are the head of a New Man.
- A New Man is made of Ear Factor, Heartwood Factor, Fire Factor, Mary Factor, Red Blood Factor, and Jewish Brain.
- The Ear Factor is Yoda.
- The Heartwood Factor is the Wookiee.
- The Fire Factor is Han Solo.
- The Mary Factor is the virgin birth.
- The Red Blood Factor is the severed "arms" of Leia.
- The sacrificial lamb is Luke.
- Luke is betrayed by Lando Calrissian, ruler of the Cloud City of Bespin.
- The Jewish Brain is Ben Kenobi.
- The Underworld is the court of Jabba the Hutt, the source of pitchblende.
- The "head" of the man-made leukoctye is transformed by (bugs' eye) lenses.

For extra credit:
Construct your artifact using only techniques that were publicly available in 1976.

THE COLOR GREEN

The absolute chronology is where they buried the dog.
Helga Wild[1]

The concept of the symbolic body changes our understanding of science and technology. Both scientists and technocrats are agreed that the essence of industrial technology is the application of scientific knowledge to the construction of artifacts and the management of natural processes. That is, scientific research when combined with industrial R&D produces both technological spin-offs and cumulative knowledge of nature. Advocates of this process point to penicillin and atomic bombs as evidence of its efficacy, but symbolic anthropology indicates that this widespread view of technological development is deficient in critical respects. First of all, this theory takes no account of the symbolism and myth that are essential in motivating the participants and enlisting social cooperation among the groups involved. Second, the folk theory maintains that the output of this technocratic process is increased "control over nature," whereas symbolic anthropology indicates that its function is to isolate us from nature, through barriers and prostheses. Third, the folk theory would have us believe that knowledge is cumulative and objective, whereas symbolic anthropology indicates that science is a process of selection, not acquisition, and that for every truth revealed by a technique, there are many others that have been systematically excluded.

Anthropology does not dispute the factuality of scientific knowledge nor deny the physical existence of cosmic rays, isotopes, genes, atoms, or whatever. To the contrary, it recognizes that the effectiveness of the technocratic system is based on the fact that every term offered to the public is backed by a host of measurements, pictures, and chemical reactions that attest to its substantiality. Also, anthropology agrees that on a superficial level the system works exactly as the scientists say it does: a particular procedure when applied in a particular context produces a certain observable effect that can be "replicated" by any other observer who reproduces the conditions. But an informed concept of knowledge, which scientists themselves accept, maintains that facts alone have little value, for they need to be integrated into a logically consistent theory with predictive and explanatory power. Not only does the theoretical context constrain the recognition of facts, the acceptance of data, and the development of new techniques of observation, as discussed in Chapter 4; but, more importantly, there are unconscious symbolic processes at work that have not been generally recognized. First, science is in structural opposition to Judeo-Christian theology, so it selects for "images of nature" that reinforce its institutional position as the successor to the church and heir to theology. Second, it is permeated with an androcentric mythology that equates knowledge with the process of transforming terrestrial nature into images of celestial light by means of lenses and a phallic apparatus. These unconscious selective processes constrain the kinds of apparatus that are built, the observations that are made, the data that are considered acceptable, and the descriptions that are presented in journals and textbooks as examples of scientific fact. Moreover, as we saw with the analysis of popular mythology, taboo symbolic content reorganizes the surface structure, modifying names and transforming imagery to make them more congruent with the repressed material, which in the case of science results in a logically consistent symbolic system that "just happens" to be contemptuous of nature and hostile to traditional religion.

Thus, science exemplifies what Gaye Tuchman, in her analysis of TV news, calls a *web of facticity:* a symbolic world created entirely out of facts which narrowly defines the boundaries of reality, presents a coherent symbolic framework for the interpretation of new information, and preempts intrusion by any other modes of discourse.[2] Even though every single statement presented in a scientific textbook or on the TV news is meticulously documented with photographs and eyewitnesses, the factuality of the statement is subordinate to the symbolic context that organizes the information. In such circumstances, the informed viewer must ask Why this fact and not another? Why this sequence of events and not the reverse? Because science is a web of facticity that assimilates more and more facts to a symbolic framework of interpretation defined by the imagery of "male" bodies and celestial light, it is possible for every single one of its facts to be true even though the picture of nature that it presents is fundamentally misleading.

But surely if science's picture of nature were false and its folk model of technological development fundamentally deficient, then it could never have produced penicillin and atomic bombs? To the contrary, the process of scientific R&D can produce any artifact or technique that is consistent with its unconscious imagery and covert agenda, such as destroying enemy aliens, separating the head from the body, making artificial starbursts, and replacing living things by their media images. But it cannot produce any technology that requires the integration of human life with the ecological processes of the planet—for this sort of integration is precluded by the underlying imagery and myth. Moreover, any logically coherent web of facticity that contains physically derived propositions can produce technological spin-offs, but the artifacts so developed in no way prove the truth of the intellectual system as a whole. The efficacy of the atomic bomb no more proves the truth of physics as a system of thought than the invention of gunpowder by Chinese alchemists proves the truth of Taoism.[3]

But surely the science of ecology makes possible the integration of science with terrestrial cycles? To the contrary, ecological science emphasizes reductionist, managerial theories based on the mathematical modeling of fragments of data wrenched from the organic context. And as shown in Exhibit 7.1, it has elevated the process of extermination to an academic methodology.[4]

Exhibit 7.1 Ecology at Harvard

E. O. Wilson, a professor of biology at Harvard, coauthored one of the canonical works in modern ecology with Robert H. MacArthur, called *The Theory of Island Biogeography*, which was published in 1967. Wilson and his colleagues chose an island in the Florida Keys, fumigated it until all the insects died, and then observed how species reoccupied the island.[5]

In American mythology, "MacArthur" is already linked to island geography through General MacArthur's "island-hopping conquest" of the Pacific; and the extermination of life on tropical islands is chartered by the Father of the H-bomb himself. Significantly, the scientific acceptance of Wilson's "ecological" theory is contemporaneous with the American defoliation of Viet Nam.

Without a fundamental change in the symbolic body, even environmentalist concerns can be easily incorporated into the technocratic worldview and reduced to another one-two punch. In this flip-flop of industrialism, shown in Figure 7.1, technology and pollution are defined as part of the Past, whereas Nature, renewed and purified, is conceptualized as part of the Future. This updated version of Nebuchadnezzar's Dream allows the technocracy to abandon heavy industry, jettison efficiency, provide incentives for solar power, and give the bones back to the Indians—while leaving the dynamic of repression unchanged. The new enthusiasm for ecological values by huge corporations, epitomized by Earth Day 1990, attests to the extent to which the environmentalist movement has been assimilated into the industrial mythology of purity and pollution.[6]

These developments underscore the fact that the pathologies of the modern world are projections of the technocratic symbolic body and not simply a collection of unrelated symptoms. The bifurcation of the globe into rich and poor countries on the basis of putative stages of technological development is vintage Nebuchadnezzar's Dream. The arms race and the threat of nuclear war are the institutional elaboration of bloody stumps and exploding stars. Global pollution, habitat destruction, and the disruption of natural cycles are implicit in the industrial imagery of nature as an alien and untamed force that needs to be made productive by the application of technique. The replacement of living things by machines is implicit in the Galilean circuit, which

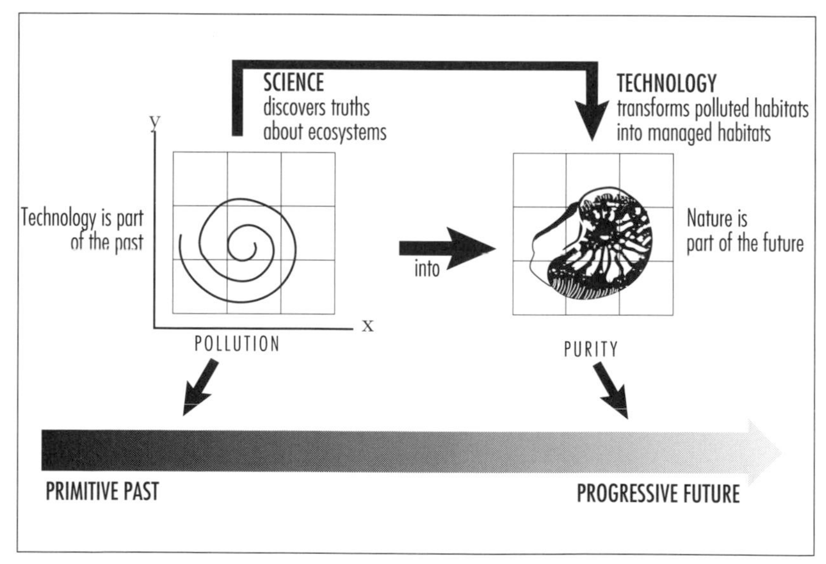

Figure 7.1 Ecology is easily assimilated to the technocratic agenda.

engenders the attitude that media images of nature have more reality than nature itself. In short, the social and ecological conditions that threaten the health and security of contemporary societies are not fortuitous events that could never have been anticipated but the logical consequence of the technocratic image of the human species and its relationship to nature.

It follows that many of the problems attributed to industrial technology are created at the level of the technocratic symbolic body and are not primarily "technological" at all. *Technology* is a universal function of human society, but the *technocracy* is a culturally specific system of myth and ritual, developed and disseminated in Europe and North America from the 17th to the 21st centuries. Even if defined in strictly modern terms, as the use of mathematics and codified physical laws in the intentional development of machines, then "technology" precedes industrialism, having served such disparate masters as Greek city states, the medieval Church, and imperial China. Conversely, *technocracy,* defined as a political ideology based on the myths of The Lone Galileo and Nebuchadnezzar's Dream, can coexist with radically different tools and techniques, as indeed it has at different periods of its history. Thus, the talkshow anxiety about "run-away technology" diverts people from the real issues, for it mistakes the symptoms for the disease, confusing technical processes and cooperative production with the symbolic structures in which they are embedded. If tools and techniques are landscape projections of the symbolic body, and if the symbolic body is chartered by myths, then there is no point in countering technology directly at all, for to do so reinforces the ideology of physical causality, objective evidence, and the supposed social significance of technical change.

But if technology is not a villain, neither is it a savior. Technical processes reflect the cosmogonic and eschatological premises of the society that organizes them; and in the present institutional context, technological development, no matter how "ecological" or "humane," is more likely to propagate the values and social organization of the technocracy itself. When the courts adopt DNA "fingerprinting," for example, to identify biological evidence left at the scene of a crime, such as blood or semen, they create a need for more geneticists, more laboratories, more technicians, more research, more safeguards, more use of science itself as the major determinant of truth or falsity. In fact, as shown by the example of Warm Springs Dam and its salmon hatchery, the more rational, practical, and democratic the proposed solution, the more likely it is to make things worse, for "rationality" has become equivalent to the uncritical acceptance of technocratic premises and "solution" a mandate for industrial R & D. Thus, to adopt a technical solution is to buy a whole institutionalized package of mechanistic models, objective measurements, standardized tests, certified experts—and if you buy now, you will receive with your purchase, absolutely free, a dedicated regulatory agency!

This analysis also reveals the futility of most middle-class reform movements, which address the technological consequences of the symbolic body while ignoring the premises that make technocratic solutions appear rational and progressive. As in the case of Warm Springs Dam, both the mutilations of nature and the proposed solutions to it are often manifestations of a single industrial process—the one-two punch; and most pragmatic programs of institutional reform either unintentionally perpetuate the underlying premises of the technocratic worldview or are themselves materialist formulations of what are essentially spiritual questions about the nature of human communities, the relationship of society to biological processes, and the goals of human life itself. In the same vein, most contemporary political debate is an argument as to which form of polity provides the most congenial conditions for further industrial growth and development. The radioactive cloud from Chernobyl was produced by a state-owned bureaucracy applying the principles of Marxist-Leninist dialectical materialism, whereas the oil spill in Prince William Sound was produced by a privately capitalized bureaucracy exemplifying the profit motive—but as all these ideologies agree on the transformation of nature by machines, it is not surprising that they all have similar effects on the planet on which we live.

In the words of Barbara Katz Rothman: "The critique of technological society is not an indiscriminate throwing away of the tools. It is an objection to the notion of the world as a machine, the body as a machine, everything subject to hierarchical control, the world, ourselves, our bodies and our souls, ourselves and our children, divided, systematized, reduced."[7] Similarly, as authors such as Berman, Merchant, and Easlea have emphasized, the problem is not technology but the metaphysics of objectivity and alienation that emerged in the 17th century, which needs to be replaced by a more holistic way of thinking in which the observer is conceptually part of the system observed.[8] The theory of the symbolic body reinforces this critique, for it demonstrates that a repudiation of the premises of the technocratic worldview is also the simplest and most practical way of dealing with the problems of modern technology—for to change the origin myths is to change their landscape projections.

The Greater the Theory, the Larger the Shadow

The theory of the symbolic body changes our understanding of science as well. In the folk model of industrial society, knowledge is both cumulative and objective, made possible by the progressive development of techniques for the scientific observation of nature. But symbolic anthropology indicates that the more "developed" a scientific theory, the more out of touch with "nature" it becomes. The scientific method is based on the assimilation of natural processes to culturally constructed categories of purity and pollution, and the phenomena classified by the folk system as terrestrial and polluting are either ignored by scientific theory or are incorporated as the "raw nature" that needs to be transformed by industrial technology. At any given point in scientific development, potential facts radiate out in all directions, so the assimilation of phenomena to theory entails both the discarding of facts that cannot be explained and the willingness to forego avenues of research that do not "look promising." Thus, for every path followed up, there is a multiplicity of other paths that have been selected out by both the formal framework of the theory and the unconscious needs of the symbolic system. Far from being a process of description, science is a process of editing, in which some of the best footage must necessarily be left on the cutting room floor. For every fact that is validated scientifically, which is to say for every phenomenon that has been successfully transformed into celestial imagery, there are other phenomena that have been *systematically excluded because they contradict the symbolic premises and logical coherence of the system.*

Consequently, "scientific progress," far from producing more and more knowledge of nature as the enterprise goes on, actually produces cultural constructions that are more and more isolated from nature, save for those few instruments of observation that are mandated by the theory itself. Similarly, the more "developed" a scientific theory, that is to say, the more facts that a theory "explains" and the more tools and techniques it charters, the more observations that have been excluded because they cannot be incorporated into the rituals of reduction. In short, every "advance" in scientific knowledge is created at the expense of terrestrial processes that refuse to be transformed into the imagery of starlight. To use a mining analogy, the more gold bricks that are stacked in the vault, the more ore that has been processed, and the more ore that has been processed, the more tailings that have been left behind. Science, like industrialism itself, requires us to focus on the gold while denying the tailings. For every "advance" in scientific knowledge, there is a corresponding increment to the shadow world of excluded imagery. Consequently, scientific knowledge is not merely provisional and incomplete, as scientists would be the first to admit, but it necessarily excludes those facts and images deemed polluting by the theory itself. The greater the theory, the bigger its shadow.

This analysis explains an otherwise puzzling fact about science, namely, that the magnitude of a scientific discovery appears to have little to do with the character or depth of the scientist who discovers it. The classic instance of this phenomenon is the discovery of the molecular structure of DNA. Although it is acknowledged by almost every commentator as one of the most important discoveries in the history of science and is touted by its own promoters as the long-sought Secret of Life, in his account of this achievement in his memoir, *The Double Helix*, Watson pioneers the image of the great scientist as unscrupulous, ambitious, callous, and self-centered—denigrating his colleagues, trivializing women, and admitting to unauthorized peeks at other people's data—while disclaiming any motivation but a lust to win the Nobel Prize.[9] Moreover, this self-serving youth of twenty-five is portrayed as largely ignorant of chemistry and disinterested in most of biology, yet he discovers the molecular structure of the magic molecule that controls the heredity of every living thing, largely by talking with the equally ambitious and egotistical physicist Francis Crick. This achievement, I confess, has always puzzled me, for it is like playing the violin perfectly the first time one picks it up; and if true, it implies that the key to life on earth is a terribly simple thing that just hangs there like a ripe plum waiting to be plucked by the first pair of graduate students with access to Rosalind Franklin's pictures. But perhaps this self-portrait of Watson is a literary device—perhaps he is really a sensitive, well-rounded human being of great philosophical depth, who heads up the Human Genome Initiative out of a sense of social responsibility and a commitment to the future of medicine? Watson then would be the sort of complex and educated intellect who deserves to discover the Secret of Life, but this hypothesis is pure speculation without support in the literature.[10]

So taking Watson at his word, symbolic anthropology suggests another interpretation: that Watson and Crick's contribution to science is a symbolic image, a logo, that crystallizes a preexisting web of facticity. That is, the Player and the Mechanist did not discover the Secret of Life at all but created a symbol, the Double Helix, that covertly referenced both the A-bombs of Hiroshima and Nagasaki and the caduceus of the medical profession.[11] Because a whole generation of biologists and medical researchers had been for more than twenty years selecting data that emulated the imagery of nuclear physics, using X-rays and radioactive probes, then a condensed image of these unconscious selection processes would, of course, "explain" almost all the data of molecular biology. Moreover, a lack of experience with the world of nature would hardly be a liability if nature were largely irrelevant to the enterprise. And if data are limited to the sense organs that the theory creates, then it is hardly surprising that every new technique "confirms" the theoretical framework, inspiring another round of mutilation and prosthesis. Thus, the easy pickings of DNA, as well as the subsequent phenomenal "success" of this model in spinning off techniques unique in their alienation, simply demonstrates the extent to which this branch of science has cut itself off from terrestrial processes. In short, the theory of life in which the facts of DNA are embedded is not a picture of nature at all but a projection of the unconscious cultural premises that frame the reductionist agenda of molecular biology.[12]

In summary, symbolic anthropology challenges the technocratic system of knowledge and thought. If objectivity produces nodes for webs of facticity, if scientific theories are the surface structure of rituals of purity and pollution, if the intellectual context is defined by the charter mythology and not by the mathematical models, if overt theory necessarily creates a covert shadow of unassimilated facts, and if tools and techniques are projections of a symbolic body collectively constructed by the culture, then it follows that *any scientific theory can be shown by anthropology to be a system of folk belief whose content is determined by its structural role in the symbolic body image.*

Big Time

However, most anthropologists in the United States, like other academics, owe their intellectual positions to the uncritical acceptance of their native culture, teaching that human evolution is a function of climate and calories, tools and techniques; and every introductory textbook in anthropology begins with a blatant display of technocratic premises: a chart that shows the position of every key site and fossil of human evolution on a single time scale as determined by physics. In textbook anthropology, the grid of time and space as defined by the physical sciences is the major metaphor for organizing human societies into a coherent intellectual framework.

This uniform time scale, from the creation of the universe to the present day, is called the "absolute chronology," and it is based on the rate of decay of radioactive isotopes as determined by the quantum theory. For example, physics tells us that radioactive carbon, C-14, decays into stable, nonradioactive carbon, C-12, at a known rate called the half-life of the isotope. A certain amount of C-14 is continually being formed by cosmic rays in the upper atmosphere, and this is incorporated into living things along with C-12, the two isotopes being present in a fixed proportion. When the organism dies, the C-14 begins to decay into C-12 at a known rate, eventually disappearing entirely. To determine the age of a sample of organic matter, such as charcoal from a fire pit, scientists measure the proportion of C-14 to C-12 remaining in the sample, then compute the amount of time that must have elapsed using the half-life value provided by physics. The radiocarbon method can be used to date things that are less than 70,000 years old, beyond which the method becomes inaccurate.

Potassium-argon dating works in a similar way, but it is used to date much older samples of volcanic rock, two million years or more.[13] The K/Ar method is based on the fact that the chemical element K-40, a common constituent of lava, changes into the inert gas Ar-40 at a known rate. When the liquid lava solidifies, the potassium it contains is trapped inside crystals, where it remains, slowly changing into argon over a period of millions of years. Consequently, if the crystals are melted in the laboratory and the ratio of potassium to argon measured, the age of the sample can be computed from the relative amounts of the two elements.

Although the textbooks present radiometric dating as an objective methodology which produces dates that are independent of culture and observer, symbolic anthropology interprets these methods as primarily expressive acts. Radiometric dating is expressive, first, of a hierarchical relationship between archaeology and physics, and, second, of the relationship between physics and the planet earth. This interpretation is supported by the social facts of radiometric dating. Archaeologists who need to have an excavation dated send portions of the materials unearthed at the site, such as pieces of charcoal and fragments of textiles, to a physical laboratory. Next, this literally terrestrial matter is consumed by fire in a sealed oven and transformed into meter readings based on a theory of starlight. In exchange, the archaeologists' findings are "validated by science," that is, assimilated into a worldview created by physics. Notice, however, that the social exchange process is highly asymmetrical, for physicists never ask archaeologists to evaluate physical observations in the light of human history and cultural variation—for that would imply that the questions and procedures developed by physicists are to some extent bound by language and tradition and are not culture-free creations communing directly with the essence of nature. But as archaeologists are as materialistic as physicists, and generally far more literal-minded, they accept this asymmetrical relationship, for it validates their own cosmology and transforms them into hard-nosed methodologists.

Radiometric dating is also a ritual expression of the relationship of physics to the planet earth. C-14 was first discovered in the late 1930's in the course of a comprehensive program to measure the decay products of atomic nuclei bombarded in the cyclotron at Berkeley, and many of the scientists were later involved with the Manhattan Project. Willard Libby, a professor of chemistry at Berkeley who publicized the radiocarbon dating method, for which he won the Nobel prize in 1960, was an outspoken booster of atomic weapons and nuclear power.[14] He served as a commissioner of the Atomic Energy Commission in its heyday of unrestricted atmospheric testing and suppressed medical statistics, and if archaeologists know him as the "discoverer of radiocarbon dating," nuclear activists know him as the man who dismissed the problem of radioactive contamination of Las Vegas with the words: "People have got to learn to live with the facts of life, and part of the facts of life are fallout."[15] He was also an enthusiastic advocate of Project Plowshare, a U.S. government program to create exemplars of progress, such as a sea-level canal in Panama, by setting off chains of nuclear explosions (Exhibit 7.2). In short, the creator of radiocarbon dating was the epitome of the modern technocrat riding roughshod over nature—and the same imagery is exemplified by the technique itself, which is a symbolic expression of contempt for the planet earth.

According to Johannes Fabian in *Time and the Other: How Anthropology Makes Its Object*, the function of radiometric dating is to replace the "absolute chronology" of sacred history with an absolute chronology based on physics; and this interpretation is consistent with both the atheological agenda and the logical necessity of a single time line to demarcate the segments of historical progress.[16] From this perspective, the role of radiometric dating is to obscure the present, not clarify the past. By the same reasoning, archaeology's other perennial theme, the stages of human history, sustains the concept of historical time as movement along the length of a body, from foot to head in the theory

Exhibit 7.2 The intellectual context of carbon-14.

Australia is populated only on its fringes. The vast, arid interior of this continent is susceptible to grand concepts that would open it up to civilization. To illustrate, Australia possesses huge quantities of iron ore in its Hamersley Range, but there is no permanent water supply to support industry or people in the area. The Fortescue River, which flows through the area, is swollen with a half-million acre-feet of water a brief part of the year, but it is usually dry. An earthen dam could be created by caving in the sides of a gorge near the Hamersley Range with nuclear explosives. An iron industry could then be built around the water reservoir and hydroelectric power source that results. Unfortunately, earthen dams eventually succumb to seepage, leaks, and general deterioration. Athelstan F. Spilhaus has suggested a neat circumvention: Inject water into the loose debris comprising the dam and freeze it, forming a permafrost barrier to seepage. After the initial freezing only a little power from the new hydroelectric plant would suffice to keep the earth-insulated ice frozen. The energy generated by a contained, underground nuclear explosion could be tapped to power the initial freezing of the dam. By pumping water into the hot cavity, sufficient steam could be generated to drive turbogenerators that would run a temporary refrigeration plant.

Glenn T. Seaborg, the discoverer of plutonium and former chairman of the AEC, discusses Project Plowshare. Note that the name "Project Plowshare" is a biblical allusion—to Isaiah.

From *Man and Atom* by Glenn T. Seaborg and William R. Corliss. Copyright © 1971 by Glenn T. Seaborg and William R. Corliss. Used by permission of the publisher, Dutton, an imprint of New American Library, a division of Penguin Books USA Inc.; pp. 115–116.

of progress (or from head to foot in theories of degeneration). Thus, the so-called absolute chronology is both a rationalization of Nebuchadnezzar's Dream and the imposition of a celestial metaphor derived from physics onto terrestrial processes. The absolute chronology is the colonization of anthropology by physics.[17]

The mythic component of absolute time is transparent once one rejects the technocratic distinction between knowledge and myth. Thus, in "the story of radiocarbon," cosmic rays in the upper atmosphere, that is, *celestial radiation*, produce radioactive carbon-14, which is incorporated into living things along with carbon-12. Thus, *radioactive* carbon is indicative of *living* things, for after death, the decay of C-14 is not counterbalanced by incorporation, and the pool of the radioactive isotope begins to disappear until only the nonradioactive (noncelestial) carbon remains. By an inverse process of transformation, potassium, literally *pot ash*, traditionally made from the ash of hardwoods, decays into argon ("no energy"), a noble gas that escapes from its terrestrial prison, moving upward to become a constituent of air and closer to the source of light. As shown in Chapter 5, the half-life is both the period of time that separates an isotope from the source of life, namely the Big Bang (either man-made or natural), and also the "half of life" that is created by physics, namely the male, celestial half, from which the female half has been removed. Moreover, radiocarbon dating (*carbon*, "burned matter," cognate with English *hearth*) is used to date materials close to us in time, particularly the remains of fireplaces, whereas potassium/argon dating is a measure of volcanic activity, that is, natural fires under the earth. Thus, the two methods of anthropology's absolute chronology, K/Ar and C-14, are related like *Star Trek*'s Mr. Spock and Captain Kirk, the former half human/half vulcan, the latter unequivocally a man. Thus, to travel backwards along the road defined by the absolute chronology is to leave the familiarity of the domestic fire and to proceed through natural but still terrestrial fire to real fire, namely celestial fire—not the sun (for this is far too parochial) but to the very origin of the universe, the Big Bang itself, a cosmic thermonuclear blast. Imagine, children, a Trinity test that generates all time and space!

Geocentric Cosmology

The history of physical science, from Copernicus on, has been the use of models of nature based on celestial phenomena, especially the imagery of light and fire, to organize matter here on earth, while systematically ignoring the human and ecological consequences of the method. What the academy calls intellectual progress is really a process of colonization in which concepts appropriate to one area of human experience are arbitrarily ruled invalid and replaced by the mathematics of celestial objects. It follows from this symbolic analysis that radiometric dating, far from being a neutral tool for producing culture-free dates, is really the assimilation of anthropology and the earth sciences to the imagery of disembodied heads and exploding stars.

The theory of the symbolic body indicates that the dates so generated, far from grounding cultural events in the eternal verities of physics, instead reveal that the chronologies of physics are themselves culturally specific systems, with no more claims to universality than perspective drawing. Once again, every fact proferred by physics is scrupulously correct, but the net effect of these truths taken together is a profound distortion of terrestrial nature and the human relationship to it. Specifically, in the clocks provided by physics, there are all manner of fires, including big bangs, solar cycles, and vulcanism, but where are the periodic terrestrial events, such as blood, hair, and biological fluids that are far more representative of life on the home planet?

Since it is clear from symbolic analysis that the arrow of progress is a surface structure rationalization for a segmented image of the human body (Figure 3.1), what I have called Nebuchadnezzar's Dream, then a linear arrow of cosmic evolution, demarcated into segments by radiometric dates, is really a landscape projection of the underlying body image and not a fact of nature itself. From this point of view, the "dates" of the absolute chronology are analogous to the "prophetic dates" in H. G. Wells's fictional history of 20th-century science. In spite of their superficial appearance to a calendrical chronology, they are not dates in the history of the earth at all but rather numerons—symbolic numbers—that mark the segmented boundaries of a culturally constructed symbolic body (Figure 7.2).

In the technocratic theory of science, the scientist stands outside of nature, observing it with mathematically constructed instruments that yield measurements independent of cultural context.[18] Because science, in this view, is an objective process of knowledge acquisition, technocrats are skeptical of a systematic relationship between the content of science and the semantic structure of language, and they reject outright the idea that taboo symbolic content is an intrinsic component of scientific activity. Symbolic anthropology, in contrast, asserts that the folk theory of science is only a small fraction of the total symbolic system, and that the terms in physical theory, although connected by mathematical relationships constructed by scientists, are also connected by symbolic associations that reflect the semantics of language and the charter mythology of scientific institutions. Where physics sees itself as a culture-free system, symbolic anthropology asserts that physics, like any other collective human activity, is necessarily mediated by culturally transmitted symbolic structures that do not meet science's requirements for a formal theory.

These competing interpretations, the anthropological and the technocratic, make very different predictions about the nature of scientific activity and its relationship to the physical world. The historical uniqueness of physics, and the source of its power and flexibility, lies in the fact that every single one of its operative symbols is connected to the world of physical reactions as well as to the world of symbolic forms. In the viewpoint of anthropology, these symbolic representations of physical processes are also embedded within a larger framework, the symbolic body. As shown in Figure 7.2, the technocratic symbolic body is a rigorously logical construction, however bizarre its internal imagery; and both its conscious and unconscious parts, in spite of their being kept rigorously distinct through rituals of purification and taboo, are nonetheless reciprocally related through symbolic operations. In science, "pure research" transforms portions of the "natural body" into starlight, and these purified materials in turn are used to construct man-made body parts which fill the holes created in the natural prototype. But as the tailings of this process cannot be overtly acknowledged in physical theory, pure research also creates a shadow world of covert symbolic forms. As shown in Figure 7.2, both the overt and covert symbolic processes of science are reciprocally related within the larger structure of the symbolic body, so it follows that the physical processes denoted by these symbols are reciprocally related as well. Consequently, anthropology predicts that the overt system of physics is reciprocally connected to a shadow world of physical processes that cannot be acknowledged by a mythology of sky gods but which nonetheless can be revealed through its *semantic* associations to menstrual blood, pubic hair, and sexual fluids—for it is these semantic relationships that caused these physical processes to be excluded from physical theory in the first place. However, in the technocratic interpretation of science, no systematic relationship between the content of physical theory and the unconscious semantic categories of natural language would be expected at all. So the two theories, the anthropological and the technocratic, provide a clear intellectual choice that can, in principle, be resolved in anthropology's favor by scientific methods alone. Moreover, this geocentric physics, if seriously pursued, would reveal aspects of the physical world without parallel in the imagery of exploding stars, proving even to mechanists that the dream of absolute chronology is not only symbolically suspect but physically impossible to attain.

The Color Green

Under the present intellectual regime, the most characteristic feature of human beings, their propensity to live in symbolic worlds, is deemed to be unnatural and must be explained away as the random permutations of atoms and economic forces. But the theory of the symbolic body argues that symbolic understanding resonates with the fabric of reality, and that the ability to make symbolic distinctions is as valid a source of knowledge and truth as mathematics, logic, and experimentation. Culturally elaborated symbolic categories, such as earth and heaven, water and fire, reflect real processes in the interactions between the human intellect and the objects of its reflection. The concepts and methods of physics, however appropriate to motions of celestial bodies and to practical problems of mechanics, when applied uncritically as a general theory of human action lead inexorably to alienation, pathology, and totalitarian institutions. When the intellectual elaborations of one symbolic domain are applied to another, they must be taught, as Cardinal Bellarmine warned Galileo, as purely mathematical constructions—not as true descriptions of nature.

The theory of the symbolic body is a radically different approach to human history and action, to technology and science, and it has implications for all fields of descriptive knowledge developed within the modern tradition. It also presents an effective challenge to the mythology of industrialism by integrating both factual and symbolic content into a coherent intellectual framework in which the cyclical processes of Mother Earth have parity with voyages to the stars. Symbolic anthropology, by making possible a rapprochement between mythical constructs and intellectual skills, can begin to heal the rift in Western culture that began with the trial of The Lone Galileo.

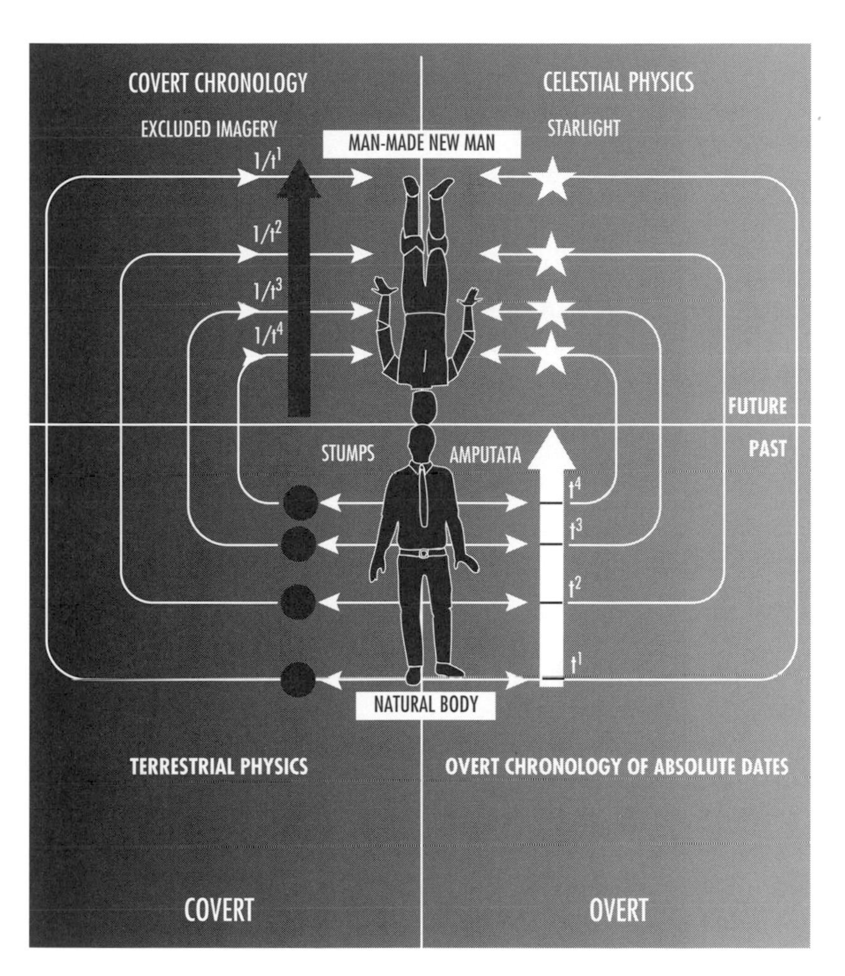

Figure 7.2 Scientific facts and covert imagery are reciprocally related within the symbolic body.

NOTES

Pumping Salmon

1. Nicole Sault, personal communication.
2. These and other environmental aspects of medieval technology are described by Gimpel, *The Medieval Machine*, 1976.
3. In maps from the Cold War years, Chemnitz is shown as Karl Marx Stadt in East Germany, but it changed its name back in 1990.
4. Ansley and Barnum, "Brave New Biotech" (Technocratic Genre Works).
5. These Du Pont ads aired on American network television in 1990. Du Pont was also a major protagonist in the development of the atomic bomb, building and managing the plants in Washington State that produced the plutonium.
6. Although Ellul's book was first published in French in 1954 as *La Technique ou L'enjeu du Siècle*, it was published in English as *The Technological Society*—obscuring the distinctions that Ellul was making.
7. The special status of trout and salmon is as old as the Indo-European language family. See the meticulous reconstruction by Diebold, "The evolution of Indo-European nomenclature for Salmonid fish: the case of 'huchen' (*Hucho spp.*)" (1985).
8. Malinowski's most accessible work is *Argonauts of the Western Pacific*, but he wrote a number of books on the Trobriand Islands based on his experiences there. In fact, he is important enough to have been deconstructed—the ultimate contemporary honor—by Clifford in *The Predicament of Culture*.
9. An overview of technology from an anthropological perspective is Forde's *Habitat, Economy, and Society*, first published in the 1930's. Although it has a strong environmental-determinist streak, it is anthropological in emphasis: each tradition of tools and techniques is presented as part of a larger cultural context that is integrated both with the habitat and with a culturally specific system of exchange.
10. Tikopian ritual is described in Firth's *The Work of the Gods in Tikopia*, but it should be read in conjunction with his treatise on Tikopian economic organization, *Primitive Polynesian Economy*.
11. In Western star charts, the Pleiades are on the shoulder of Taurus the bull.
12. For examples, see Joseph Needham on Chinese iron-working.
13. The books by Mary Douglas are a good introduction to the anthropological perspective on myth and ritual. The structuralist study of myth is developed at length in the works of Lévi-Strauss and Edmund Leach. Leach has also written a clear introduction to Lévi-Strauss's thought for the general

reader. Key concepts from the Prague school of phonology, from which structuralism developed, are used in the book by Jakobson and Halle. In the 1950's and '60's, anthropologists applied the methods of structural linguistics to systems of folk classification, a process described in the articles by Frake (1962) and Conklin (1955); and Harold C. Conklin's bibliography, published in 1972, lists more than 5,000 works relevant to folk taxonomies including Lounsbury's. For a contemporary account of semantic categories and their relation to cultural systems, see the book by Lakoff (1987).

14 Knowledge that people use but cannot articulate is commonly referred to as *tacit knowledge* by anthropologists, whereas the "unconscious" is more often reserved for the psychological theory of repression, in which mental content is not only tacit but made so by the person's own defense mechanisms. As I employ repression as a theoretical construct, I do not try to distinguish between tacit and unconcious knowledge but use the term *unconscious* to refer to both.

15 For a readable introduction to the English language based on linguistic analysis, see the books by Pyles (1971) and Williams (1975).

16 Elsewhere, I have shown how to analyze complex technical skills using video techniques and linguistic concepts of structure (Reynolds, in press, and the references cited therein).

17 Although social anthropologists have long known that the image of the primitive is a cultural construction of European scholars, this insight has only become intellectually fashionable in the last decade, primarily as a result of works with a Third World perspective, such as Said's *Orientalism.* Unfortunately, the cover of the Vintage paperback edition (1979) of Said's book, with its painting of a naked Oriental snake charmer, panders to exactly the sort of eroticized exoticism which is debunked in the text.

Nebuchadnezzar's Dream

1 As the space shuttle *Atlantis* indicates, the name is still symbolically important in technocratic culture.

2 Wallace's *The Social Context of Innovation* documents the relationship between industrialism and the biblical Book of Daniel. See also Bacon's collected writings in the bibliography on Technocratic Genre Works.

3 For biblical texts I have used *The New English Bible with the Apocrypha: Oxford Study Edition* (Sandmel et al., 1976).

4 Daniel 2:1; the dream itself is in 2: 31–36.

5 Both Hesiod and Ovid are translated in Hendricks (1974).

6 See Daniel's *The Origins and Growth of Archaeology*, Chaps. 4 and 5, as well as Glob's very readable works on Danish antiquities.

7 Morgan's *Ancient Society*, where the stage theory is explicated, is listed in the bibliography on Technocratic Genre Works.

8 The quote from Childe is given by Glyn Daniel in *The First Civilizations,* p. 143.

9 For an important critique of Morgan's description of Indians by a cultural historian, see the chapter "Montezuma's Dinner" in Keen's *The Aztec Image*

in Western Thought. Also, Morgan's interest in Indians was an outgrowth of his lifelong participation in male cults that appropriated Indian symbolism. This little-known aspect of Morgan's life is well documented by Carnes (1989, pp. 93 ff.).

10 The remark is by Fiske, author of the *The Discovery of America* (1892), and it is quoted by Keen, p. 391. For a scale model of the Aztec capital at the time of the Spanish conquest, see the illustrations in the article by Alfred Meyer (1980). An overview of New World metallurgy is given in Clark's *World Prehistory.*

11 The paper by van der Merwe and Avery (1982), "Pathways to Steel," is an excellent introduction to the diversity of metallurgical traditions in the Old World.The book edited by Wertime and Muhly about prehistoric iron metallurgy, *The Coming of the Age of Iron*, is more complete but also more technical. Aitchison's *A History of Metals* is worth looking at for the illustrations of archaeological finds, but the text is uncompromisingly Eurocentric and progressivist. Crawford's *Subterranean Britain* has fascinating material on prehistoric mining. Also, Gimbutas's *Bronze Age Cultures in Central and Eastern Europe*,with its beautiful illustrations of archaeological finds, should dispel any notions of "primitive" craftsmanship. For the Chinese contribution and its relationship to European industrialism, Joseph Needham's works are essential. Unfortunately, as Needham's findings do not support the official ideology of Eurocentric progressivism, they are almost impossible to get hold of except by professional scholars with access to university libraries.

12 For the Tasmanians, I have used the work of Peter White and his colleagues, as well as Flood's *Archaeology of the Dream Time.*

13 Sandra Bowdler.

14 Clark and Piggott (1965) describe the aftermath of the Upper Paleolithic in Eurasia.

15 Unfortunately, I do not know of a single, comprehensive work that describes Aboriginal societies in a way that is understandable to a general reader, portrays the cultures accurately, and takes account of postindustrial anthropological perspectives.

16 New Guinea is the name of the island, which is now divided into two political entities, namely, the independent country of Papua New Guinea and the Indonesian province of Irian Jaya.

17 This and subsequent quotations are from Strathern's *Ongka: A Self-Account by a New Guinea Big-Man.*

18 Strathern, p. 2.

19 Strathern, p. 8.

20 The efficacy of stone tools is described by Coles in *Archaeology By Experiment.* The economics of metal versus stone tools in New Guinea is mentioned by Young (1971, pp. 173, 255–256) and Salisbury (1962).

21 Summarized by Glyn Daniel, *The First Civilizations*, p. 24.

22 As an example of the archaeological use of the term, see Glyn Daniel, *The First Civilizations*, and Mellaart, *The Earliest Civilizations of the Near East.* Also, for this section I have used de Camp, *The Ancient Engineers;* Forbes, *Studies in Ancient Technology;* French, *The Story of Engineering;* Grant,

The Cities of Vesuvius; Hodges, *Technology in the Ancient World;* MacKendrick, *The Mute Stones Speak;* as well as Joseph Needham's works on China.

23 For subsistence skills in ancient Egypt, including reproductions of many of the murals, see Aldred, *The Egyptians;* Montet, *Everyday Life in Egypt;* and Wilson, *The Culture of Ancient Egypt.*

24 Forbes's *Studies in Ancient Technology* is an important source on prime movers in Greek and Roman culture. A Greek work from the Hellenistic period describing a broad range of mechanical devices, *The Pneumatics of Hero*, is available in English (Hall, 1971). Lloyd, in *Magic, Reason and Experience*, discusses the scientific theories underlying Greek machines.

25 Brian Fagan has described the vigorous ingenuity of early antiquarians in several highly readable books, such as *The Rape of the Nile* and *Elusive Treasure.* For the work of Schliemann, see Brackman, *The Dream of Troy*, as well as Glyn Daniel's books.

26 Glyn Daniel, *The Origins and Growth of Archaeology*, p. 145.

27 Daniel, ibid.

28 Ward-Perkins, 1972.

29 This description is taken from Cline, *Mining and Metallurgy in Negro Africa.* For a contemporary account that focuses on the social context of African metallurgy, see McNaughton.

30 McNaughton, *The Mande Blacksmiths.*

31 I have used Pope's treatise, *Bows and Arrows*, combined with information on the Hyksos invasion in the source books on Egyptian culture (see above). Ishi's contribution is described by Kroeber, *Ishi in Two Worlds.* For a revealing account of the role of the bow in warfare, read Keegan on the Battle of Agincourt in *The Face of Battle.*

32 See Sandars, *The Sea Peoples.*

33 Roman weapons and fortifications are very well known, from both archaeological and historical sources. The archaeological reconstructions in Parker (1984) convey the huge size of even provincial Roman fortifications. The network of walls is described by Webster (1969). MacKendrick (1960) discusses Roman construction techniques.

34 The illustrations in Soedel and Foley's article (1979) on Greek and Roman catapults are worth examining. Marsden (1969) provides extensive documentation.

35 Hall, 1971.

36 See Forbes's *Studies in Ancient Technology* and *Man the Maker*. Joseph Needham (1964, p. 7) describes ancient Chinese factories.

37 Mill statistics are given in Finn, *An Introduction to the Domesday Book.* Gimpel, *The Medieval Machine*, discusses preindustrial prime movers, as does Hunter, *Waterpower.* See also general works such as Cardwell, *Turning Points in Western Technology.*

38 See Braudel's discusson of energy sources.

39 Basalla, *The Evolution of Technology*, p. 211.

40 For the technical history of the early industrial period, I have used the works of Buchanan and his colleagues on the industrial archaeology of Britain;

Wallace's description of the development of the steam engine; Dickinson's works on Watt and his colleagues; Gillispie's (1959) reprint of the plates from the French Encyclopedia; Rolt's works on Newcomen; Triewald's (1928) firsthand description of the Newcomen engine; and the reproductions of period documents in Robinson and Musson (1969).

41 These data are from Buchanan and Watkins, *The Industrial Archaeology of the Stationary Steam Engine*, p. 124.

42 When air rushes in to fill a vacuum, it creates a low pressure region behind that can suck up water by the siphon principle—that is, the weight of the atmosphere pushes down on the water until equilibrium is reached. Torricelli showed that the effectiveness of the siphon principle is a function of the weight of the atmosphere, experimentally determining that atmospheric pressure can raise water 32 feet at sea level, less at higher elevations —so that siphons have inherent limits when used as pumps. However, if the rush of air into a vacuum is used to repetitively move a pump handle, then the amount of water lifted depends on the number of strokes, not on the weight of the atmosphere.

43 Wallace, p. 61.

44 The first commercial atmospheric pressure engine was erected as a water pump at a coal mine near Birmingham in 1712, where it raised 128 English gallons a minute from a depth of 150 feet. See Triewald's description, as well as Buchanan (1972). Although the term *fire engine* now means a pump truck that puts out fires, in Newcomen's time such devices were powered by hand.

45 See Dickinson (1939) on Smeaton.

46 See Dickinson (1936, 1937) on Watt and his partner Matthew Boulton.

47 For cultural as opposed to technical definitions of industrial technology, see Ellul, *The Technological Society;* Mumford, *Technics and Civilization;* and Romanyshyn's *Technology as Symptom and Dream*. Barrett, in *The Illusion of Technique*, explores the philosophical implications of a technocratic society. Winner, *Autonomous Technology*, deals with the political response to technocracy. Both Congdon, *Introduction to Appropriate Technology*, and Nash, *Progress As If Survival Mattered,* try to provide a technical alternative to technocracy. Florman (1976), confusing technique with technocracy, attacks Ellul for being opposed to engineering. Arthur O. Lewis's book of readings, *Of Men and Machines*, brings many of technology's critics and observers together in one volume.

48 Wallerstein, *The Modern World-System*, and Braudel, *Civilization and Capitalism*, Vol. 1.

Bloody Stumps, Exploding Stars

1 The use of visual material to infer cultural premises can be found in Bucher's *Icon and Conquest: A Structural Analysis of de Bry's "Great Voyages"*; Dijkstra's *Idols of Perversity: Fantasies of Feminine Evil in Fin-de-Siècle Culture*; Keuls's *The Reign of the Phallus* (a study of Greek vase painting); Munn's *Walbiri Iconography* (on Australian Aborigines);

Theweleit's *Male Fantasies* (a study of fascism); Tuchman's *Making News* (on television journalism); and Weart's *Nuclear Fear: A History of Images* (on the mythology of atomic energy). Marshak's *The Roots of Civilization* belongs with this tradition in terms of visual content, but the interpretation of the material is technocratic (Wow! Primitives think!). For the use of etymology, semantics, and textual exegesis to infer cultural premises, see the works of Bauschatz (*The Well and the Tree*) on early Germanic culture and Lincoln (*Myth, Cosmos, and Society*) on Indo-European creation myths. For technocratic definitions and etymologies of English words I have used *Webster's New World Dictionary* (Neufeldt and Guralnik, 1988). For Indo-European roots, I have generally referred to Watkins (1985).

2 I have used Delany's (1980) definition of sci-fi as the creation of imaginary worlds that are to be taken literally. H. J. Muller (1973, p. 22, in Technocratic Genre Works) has argued that science fiction performs an inspirational function for science, giving readers views "too daring to pass muster with the editors of less licensed publications." Many people, including Muller, would argue that true science fiction enhances the materialistic world view by extending the process of reduction to hitherto unreduced aspects of life, whereas science *fantasy* resurrects fabulous and magical elements in a scientific context. As science fiction and fantasy are increasingly merged, I take *Star Wars* to be an example of sci-fi, as does Searles (1988), who has created a beautiful coffee-table book of stills and plot summaries for both sci-fi and fantasy films. Wendland (1985) also summarizes the themes of many sci-fi plots. Dewey (1990) shows that the apocalyptic theme is not restricted to the sci-fi genre but is common in postwar American fiction.

3 The *Star Wars* films discussed in this section are listed as anonymous entries at the beginning of the annotated bibliography, Technocratic Genre Works. For the spinoff books, see Daley (1979, 1980) and Vinge (1983) in the same bibliography. For the spelling of characters' names, I have used the spinoff books and Arnold, 1980 (see References). Significantly, Arnold (*Once Upon a Galaxy: The Making of "The Empire Strikes Back*," p. 222) quotes George Lucas, the creator of *Star Wars*, as saying: "People 500 years from now will look at our films and be able to figure out what we were like. Our moods, our hopes, our dreams will be revealed to them." We don't have to wait that long.

4 Noam Chomsky made this terminology famous by giving it a technical definition within his theory of transformational grammar (*Syntactic Structures*), but the basic concept is implicit in any algorithm, computer program, or cultural rule.

5 In addition to *The Interpretation of Dreams*, the other works by Freud in the bibliography are especially pertinent, as they contain many examples of symbolic transformation in waking life.

6 In the postwar years, social psychologists used repression to explain the mass hysteria of the Nazi period (for example, Theweleit, 1989); and in the 1960's, Frantz Fanon (1970) used it to explain the social institutions of colonialism in works such as *Black Skin White Masks*. But these

psychoanalytic concepts have even more explanatory power when combined with the intrinsically social concept of the symbolic body.

7 In the caption, Searles, 1988, p. 29.

8 Coincidentally, SDI is the inverse of IDS.

9 References to *Alice in Wonderland* are common in technocratic genre works. Laurence (1959, p. 208, in Technocratic Genre Works), in a chapter entitled Alice in Thunderland, describes how a baby girl was born to a woman native to the Marshall Islands at the moment the hydrogen bomb was exploded on Eniwetok atoll. The child was named Alice after the wife of the chairman of the Atomic Energy Commission, Alice Strauss, who presented the mother with ten pigs.

10 The bikini is named for another Marshall Islands nuclear test site (1946).

11 There are two good books on AIDS readily available in paperback: Shilts's *And the Band Played On* is a first-rate description of institutional responses to AIDS in the United States, based on his work as a journalist for a San Francisco newspaper; Connor and Kingman's *The Search for the Virus* describes the scientific interpretations with clarity and precision.

12 Martin (1987, 1990), as well as a paper presented at the American Anthropological Association Annual Meeting, 1989. Similarly, Robbie Davis-Floyd, in her interviews with American mothers, documents a subculture in which people believe that their bodies are unreliable machines, prone to breakdowns, whose deficiencies must be compensated for by scientific technique, which in turn is a manifestation of mind. See Chap. 6 below.

13 AIDS as a projection of the broader cultural belief system is clearly presented in the book *Whatever Happened to Divine Grace?* by Ramón Stevens (Chap. 33). As Stevens claims his book was channeled by an entity named Alexander, I have listed it in References under A. See also Sontag, 1990.

14 A photograph of this artifact can be found in Marshak (1972).

15 Arnold, 1980, p. 199.

16 Arnold, p. 227.

17 Although the medical profession in the United States since the 1960's has paid lip service (so to speak) to breast-feeding, the presumption of bottle-feeding is still encoded in hospital practice. Based on an extensive review of pediatric textbooks from 1897 to the present, Millard (1990) concludes that the medical presentation of breast-feeding revolves around precisely timed feeding schedules. That is, heavenly bodies take precedence over biological bodies.

18 The purity and pollution code of the ancient Jews, which is relevant to this discussion, as will be shown in Chapter 6, is explicated by Countryman in *Dirt, Greed, and Sex*. See also Leach (1964) and Leach and Aycock (1983).

19 Theweleit, 1989, Vol. 2, p. 361.

20 Theweleit, 1989, Vol. 2, p. 59.

21 Nicole Sault points out to me that a subliminal androcentricity is characteristic of Lucas's other movies as well. In *American Graffiti*, for example, a film about a group of teenagers graduating from high school in a small American city, the end titles tell what happens in the future to the male characters but not to the female ones.

The Lone Galileo

1 Personal communication.
2 The works discussed in this book as exemplars of technocratic myth (Bronowski, Burke, etc.) are listed under Technocratic Genre Works. Books of history and analysis that are not primarily technocratic in nature can be found under References. The latter include Van Helden's translation of *Sidereus Nuncius* (Galileo, 1989); Redondi's re-analysis of Galileo's trial; Kuhn's concept of paradigm shift; Feyerabend's re-analysis of Kuhn; Funkenstein's analysis of the relationship of theology to physics; Easlea's critiques of the culture of physics (1980,1983); Romanyshyn on the relation between art and science; Lovell's history of astronomy; and Bergstein's observations about the relationship of quantum physics to language (even though the latter points are not developed here). I have also used Gillispie's sixteen-volume *Dictionary of Scientific Biography*, which is a treasure trove for anyone interested in the history of science, containing about five thousand essays on scientific heroes by contributing scholars.
3 Bronowski, p. 200 (Technocratic Genre Works).
4 Van Helden, in *Sidereus Nuncius*, pp. 6–7, who is "following, with minor modifications, the translation in Stillman Drake, *Galileo at Work: His Scientific Biography* (Chicago: University of Chicago Press, 1978), p. 141" (footnote 17).
5 Bronowski, pp. 202–204 (Technocratic Genre Works).
6 Preston, 1987.
7 These concepts can be found in introductory neuroanatomy texts.
8 Romanyshyn, 1989, p. 42 .
9 Romanyshyn, p. 42 .
10 Romanyshyn, pp. 74–75.
11 Romanyshyn, p. 74.
12 Romanyshyn, p. 95.
13 Bronowski, pp. 211–212 (Technocratic Genre Works).
14 Burke, p. 136 (Technocratic Genre Works).
15 Burke, p. 134 (Technocratic Genre Works).
16 See Yates for a discussion of the influence of Hermetic thought on Renaissance astronomy. Umberto Eco's novel, *Foucault's Pendulum* (1989, References), captures the flavor of hermetic cults.
17 Lovell, p. 47.
18 Ironically, few modern scientists believe that there is one, single, true mathematical description of nature; and many modern mathematicians doubt that mathematics is a description of nature at all. For an informed discussion of the nature of mathematical knowledge, see Kline, *Mathematics: The Loss of Certainty*.
19 Van Helden, in *Sidereus Nuncius*, p. 89.
20 In view of the Church's problems with Galileo's rejection of Aristotle, it is ironic that Aristotle was a newcomer to the Christian ranks. Forgotten during the Middle Ages, his works had been reintroduced to Christendom via Jewish and Arabic scholars in the 13th century.
21 Quoted by Bronowski, p. 218 (Technocratic Genre Works).

22 Bronowski, p. 349 (Technocratic Genre Works).
23 Galileo, *Sidereus Nuncius*, p. 30.
24 Ibid., pp. 30–31.
25 Hall, *California* magazine, p. 68 (Technocratic Genre Works).
26 The development of mechanisms for conferring immortality, such as the national academies, was important in the institutionalization of science, as shown by Paul, *Science and Immortality* (1980).
27 See, for example, the books by Iain Paul (1982) and H. Nebelsick (1981) (Technocratic Genre Works).
28 Nebelsick, 1981, p. 160 (Technocratic Genre Works).
29 Some readers have pointed out that there is no necessity for science to replace theology or to be reconciled with it, for many contemporary scientists are mechanists with no ax to grind with regard to theology. Conversely, there are some scientists, such as Polkinghorne (1988, 1989), who see such reconciliation as both possible and desirable. Although this may be true, the origin myth still charters a structural opposition between science and theology, and the mechanists still answer to the players. Maybe it is time that science got a new culture hero.
30 Quoted by Bronowski, p. 216.
31 Some have suggested that Sakharov is both a scientist and a hero. I maintain that his heroism about civil rights was made possible by his willingness to build H-bombs—that it is a reward for faithful service.
32 Rhodes, p. 171 (Technocratic Genre Works).
33 Although Einstein signed lots of peace petitions as a young man in Europe, he was not the pacifist he claimed to be. He was willing to participate in the Manhattan Project, but the U. S. government considered him a security risk (Regis, 1987, p. 127). Nicole Sault (personal communication) suggests that Einstein advocated the production of the atomic bomb because he thought he would get to work on it.
34 Hall, p. 75 (Technocratic Genre Works).
35 Hall, p. 158 (Technocratic Genre Works).
36 Weart, 1988, p. 6.
37 In Berman, *The Reenchantment of the World*, pp. 109–110 .
38 Toulmin, *The Return to Cosmology*, p. 24.
39 One reader, a physicist, tells me that many scientists have little interest in the concept of science as a generator of truths, for they see science as a discovery process and truth as a concern of philosophers.
40 Kuhn, p. 169.
41 Kuhn, p. 157.
42 Kuhn, p. 115.
43 I have discussed the intellectual background to Darwin's theory in my previous book, *On the Evolution of Human Behavior* (Reynolds, 1981), Chapter 1, drawing especially on the work of Lovejoy (1960). See also Gillispie (1959) and Eiseley (1961).
44 Scientists read popularizations too. McCarty, in *The Transforming Principle*, p. 20, discusses the role of books such as Paul de Kruif's *Microbe Hunters* in his choice of medical research as a career.

The Twins Who Bring The Fire

1 Personal communication.
2 I have used Rhodes's *The Making of the Atomic Bomb* (Technocratic Genre Works) as the primary source on the Manhattan Project. Whenever possible, I have confirmed his facts from other sources. For the imagery of the Manhattan Project and nuclear physics, I have used the works of Weart (1979, 1988), supplemented by field trips to the Bradbury Science Museum and the Historical Society Museum at Los Alamos, to the Jornada del Muerto in southern New Mexico, and the Lawrence Livermore Laboratory in California.
3 In March, the results of the clinical trials with penicillin were published, and a team of specialists was assembled in Peoria, Illinois, to begin high-volume, commercial production (Hare, p. 176; Williams, Chap. 4); in that same month Oppenheimer took up residence in Santa Fe, New Mexico.
4 For example, the word "penicillin" does not appear in Rhodes's index. In Jonas's history of the Rockefeller Foundation, Oppenheimer appears as an example of a famous physicist who had received a Rockefeller fellowship (Jonas, p. 305).
5 For the history of penicillin, see Macfarlane, *Howard Florey: The Making of a Great Scientist;* Williams, *Howard Florey: Penicillin and After;* and Hare, *The Birth of Penicillin and the Disarming of Microbes.*
6 For information on the Rockefeller Foundation, I have drawn primarily on Jonas, *The Circuit Riders: Rockefeller Money and the Rise of Modern Science.*
7 Jonas, p. 113.
8 Jonas, p. 174, quoting Warren Weaver.
9 Jonas, p. 150. The Baptist religion, which forms the backdrop to the culture of the Rockefeller Foundation, is a version of Christianity centered around a rite of purification. For an anthropological study of a contemporary Baptist church, see Heriot's thesis, *Blessed Assurance*, 1989.
10 Jonas, p. 151.
11 Jonas, p. 167.
12 Jonas, p. 203.
13 Jonas, p. 203.
14 Jonas, p. 206.
15 Jonas, p. 207.
16 See the references to "quantum" in Warren Weaver's memoir, *Scene of Change: A Lifetime in American Science.*
17 Jonas, pp. 186–188.
18 Jonas, p. 198.
19 Jonas, p. 226. Over the next three decades, the foundation's contributions to quantitative "vital processes" would total more than 90 million dollars (Jonas, p. 227). Of 18 Nobel laureates between the years 1954–1965 who made significant contributions to biochemical genetics, 15 had been recipients of Rockefeller grants before receiving their prizes (Jonas, p. 228).
20 Rhodes, p. 360, confirms Lawrence as a recipient of Rockefeller funds. Jonas says that he received $1,150,000 (pp. 302–303) and that the Rockefeller Foundation paid for Florey's trip to the United States. Weaver visited

Lawrence in Berkeley in February, and in April he visited Florey in London. However, neither Florey nor Lawrence are mentioned in the index of Weaver's memoir, *Scene of Change*, which in other respects as well is a remarkably uninformative document, considering the centricity of its subject.

21 The atomic bomb project was called by different names by different protagonists. The physicists at Los Alamos often referred to the bomb as "the gadget"; the army called the project the "Manhattan Engineering District"; and the Washington bureaucrats denoted the bomb and the project as "S-I," that is, Roman numeral one, after Section I of the Office of Scientific Research and Development. The term "Manhattan Project," apparently a postwar neologism, has passed into general usage.

22 Traweek, *Beamtimes and Lifetimes*, Chap. 4.

23 Traweek, Chap. 3.

24 Traweek, p. 158. This is confirmed by Carol Cohn (1987), in an article in the *Bulletin of the Atomic Scientists* describing her year's research with nuclear strategists in the Pentagon and associated think tanks.

25 Easlea, *Fathering the Unthinkable*, p. 92. Some readers have commented that physics is no different from other human institutions in this respect, but as I argue below the phallic and birth imagery is required by physics' role in the symbolic body. For the role of names in the vitalization of machines and the rationalization of content, see also Brown (1983); M. Gyi, (1984); and George Orwell (1949). Hilgartner et al. in *Nukespeak* provide an appendix of sanitized terminology.

26 This story is told by Rhodes, pp. 450 ff.

27 The eviction letter can be read in Los Alamos Historical Museum. Rhodes, however, says simply that the school was willing to sell.

28 Produced by Steven Spielberg with a story by George Lucas and Menno Meyjes (1989). The imagery of the Boy Scouts is highly appropriate to the iconic deep structure of the technocracy. As Rosenthal shows in *The Character Factory* (1986, pp. 255ff.), the Boy Scouts combined a highly racist covert ideology, exemplifed by Baden-Powell's proclamations in his capacity as an officer of the South African Constabulary, with an overt commitment to universal brotherhood expressed by the charter of the organization. Also, like the technocracy, the Boy Scouts deny any military agenda while elaborating military imagery.

29 Incidentally, *pinos,* Spanish for "pine trees," sounds like *penis* in English.

30 Rhodes, p. 121. There may be a typo in Rhodes's book, which perhaps should read "at last in shape." Significantly, the story of Oppie's boss, Henry Stimson, secretary of war under the second Roosevelt, also features a man-making stage in the Wild West, described in Hodgson, *The Colonel: The Life and Wars of Henry Stimson, 1867–1950*, pp. 38–40. This mythic scenario was first popularized by Stimson's hero, Teddy Roosevelt.

31 An exhibit in the Los Alamos Historical Museum says that the mailbox numbers were 1663 for civilians, 1559 for military.

32 Historians now have doubts about the extent of Fuchs's contribution to the Russian atomic bomb. See Broad, 1990.

33 The treatment of nuclear energy by the American popular press on the eve of World War II is described in Hilgartner et al., *Nukespeak: The Selling of Nuclear Technology in America*, pp. 17–21.
34 See Rhodes, p. 605, on the cost and volume of the atomic bomb effort.
35 Hilgartner et al. (1983, Chap. 5) show that the major function of secrecy in the postwar years was to keep the American public ignorant of government actions.
36 Rhodes, pp. 272, 274.
37 Weart, *Nuclear Fear,* 1988, p. 77.
38 See Weart (1988, pp. 87–88) for birth and generation metaphors in descriptions of the first "atomic pile," that is, in the first controlled nuclear reaction, in Chicago, 1942.
39 Weart, *Scientists in Power,* 1979, p. 67.
40 Hilgartner et al., pp. 14–17.
41 Wells, 1914, p. 110.
42 Wells, 1914, p. 112.
43 Weart, 1988, Chap. 3. Exposure to naturally radioactive elements is still considered medicinal by some, as shown in the article by Morgenthaler (1990).
44 Rhodes, p. 355.
45 The Roman Pluto, lord of the underworld, was assimilated to the Greek god Hades, but the latter is just the opposite of a god of fertility. In the prominent Greek myth of Demeter, the goddess of agriculture, the land becomes barren when Demeter's daughter Kore is kidnapped by Hades and taken to the underworld. It is only when Kore returns from the kingdom of Hades that fertility is restored (Hendricks, 1974).
46 Rhodes, pp. 350 ff., describes the discovery of neptunium.
47 Leach, 1964.
48 Seaborg, *Nuclear Milestones,* p. 12.
49 Life-size facsimiles of these bombs are on display at the Bradbury Science Museum at Los Alamos. While I was at the museum, several grammar school classes were using the exhibits to complete science class assignments. The children had been given sheets of questions to answer by their science teachers. One question was: "Why were the first atomic bombs called Fat Man and Little Boy?" A good question. So I asked one of the teachers for the answer sheet. It said: "Because one is fat and the other skinny."
50 Rhodes, p. 654, says the tower was 100 feet tall.
51 Many histories of the A-bomb do not mention Jumbo, nor is it mentioned in Rhodes's index, but photographs of it are on display at the Los Alamos Historical Museum. I am assuming that Jumbo was hung from its own tower, not the shot tower, but this is not made explicit.
52 The development of the atomic bomb was administered by a joint military and civilian committee, headed by Vannevar Bush, director of the Office of Scientific Research and Development, who reported in turn to Roosevelt's secretary of war, Henry Stimson. General Groves was the chief executive, charged with bringing the project to a successful conclusion, and he hired Oppenheimer as scientific director of the project, with responsibility for assembling the raw materials into a weapon. In Hodgson's cinematic

analogy, Groves was director, Bush producer, and Stimson executive producer (Hodgson, *The Colonel,* pp. 294–295). In popular histories of the bomb, Groves and Oppenheimer predominate, whereas Stimson and Bush are background figures.

53 In *Enola Gay*, p. 171, Thomas and Witts report that Little Boy was originally named Thin Man by Groves, a name which denoted Roosevelt, but it was changed to Little Boy when the gun barrel had to be shortened. (The Thin Man also denoted a popular series of detective movies from the period, featuring a pair of stylish, husband-and-wife sleuths played by Dick Powell and Myrna Loy.) Fat Man referred to Churchill. The atomic myth preserves the association with male leaders.

54 Foreign words and phrases can function in English mythologies provided that the translation is part of the myth. Almost every book on Los Alamos tells us what the name means. For botanical data, I have used Rushforth's *Pocket Guide to Trees.*

55 The lightning imagery is very prominent in the Trinity myth, where a fierce lightning storm delays the test from 2 A.M. to 5:30 A.M. It is also a common warfare image, as in *Blitzkrieg* and Operation Desert Storm.

56 The Trinity site was not government property before the test but was the privately owned McDonald Ranch. Its residents were evicted, and after the war, the site was incorporated into the White Sands Proving Ground "at Alamogordo." Incidentally, the sand in the area around the Trinity site is not white but red.

57 The specially constructed sites at isolated locations, namely Hanford, Oak Ridge, Los Alamos, and Trinity form an enormous but flattened triskelion (a three-spoked solar wheel) on a map of the United States. The latter image occurs in the official insignia of army personnel assigned to the Manhattan Project, which features a lightning bolt flowing from a blue, five-pointed star above to a gold triskelion below. This insignia is on display at the Los Alamos Historical Museum.

58 Rhodes, pp. 603–605.

59 The tanks were in fact much larger than the *Queen Mary.*

60 Spencer Weart discusses the relationship between radioactive waste and excrement in *Nuclear Fear,* pp. 298–299.

61 The imagery of Indian and buffalo figures prominently in the personal mythology of Henry Stimson, the nominal head of the Manhattan Project. See Hodgson (1990, pp. 23–25) for the story of Stimson's discovery of a buffalo skull on the peak of Chief Mountain. Carnes (1989) provides the cultural context for this sort of memento.

62 Rhodes, p. 427.

63 Crichton, 1980 (Technocratic Genre Works).

64 Rhodes, pp. 578 ff .

65 The number 32 is also the left/right inversion of 23, a symbolic number discussed in the next chapter.

66 The Atomic Energy Commission claims that there was no significant fallout in the first hydrogen bomb blast in 1952—but they made no attempt to measure it beyond 50 miles from the test site (Hilgartner et al., p. 98). In 1954, a subsequent blast irradiated Marshall islanders over 300 miles

away, as well as a boatload of Japanese fishermen in a "safe" area, killing one of the latter and giving radiation sickness to the others (Freeman, *Nuclear Witnesses: Insiders Speak Out*, pp. 173 ff.).

67 Oppenheimer's remark is quoted in Rhodes, p. 776. The hypothetical H-bomb is described and illustrated in Rhodes, pp. 774 ff.

68 In 1979, the United States government obtained a court order against the *The Progressive* magazine to prevent the publishing of *previously published* facts about the hydrogen bomb, declaring them retroactively classified. *The Progressive* spent over $200,000 in legal fees to contest the decision, whereas the government prosecutors used taxpayers' money to defend the government's actions (Hilgartner et al., pp. 66–71).

69 Rhodes, p. 777. It is interesting that Teller did not witness the test in the South Pacific but monitored it on a seismograph from his laboratory in Berkeley. In those days, fathers did not attend births. Teller was also referencing history. When Henry Stimson was with Truman at the Potsdam Conference in July 1945, he received the following message from the War Department confirming the results of the Trinity test: "Doctor has just returned most enthusastic and confident that the little boy is as husky as his big brother. The light in his eyes discernable from here [Washington] to Highold [Stimson's estate on Long Island] and I could have heard his screams from here to my farm" (quoted by Hodgson, p. 334).

70 Fleming's nickname is given in Hare, p. 57. *Flem* is homonymous with *phlegm*. Significantly, Fleming himself considered his most important discovery to be lysozyme, an antibiotic substance found in tears, saliva, and mucous membranes (Hare, p. 59).

71 Hare, Chap. 3.

72 See Macfarlane, pp. 282–284, on pre-penicillin antibiotics.

73 Macfarlane, pp. 279–285 and p. 238.

74 Quoted by Macfarlane, p. 285.

75 Macfarlane discusses Florey's motives, pp. 285–286.

76 Macfarlane, pp. 288–289.

77 The imagery of the berry connects penicillin to *Star Wars*—from coccus (Gk. kokkos, "berry") to bacca (Lat., "berry") to Chewbacca the Wookiee (see Chap. 3 above). Note, however, that there are no strictly etymological connections between coccus and bacca, indicating that linguistic etymologies need to be supplemented by analysis of nonverbal or iconic content.

78 The date is given in Macfarlane, p. 255. Of course, Szilard, the Jewish progenitor of the A-bomb, leaves Germany in 1933 too (Hilgartner et al., p. 14).

79 Macfarlane, p. 299.

80 The imagery of the mushroom cloud connects atomic energy to technocratic versions of the caterpillar scene in *Alice in Wonderland*, as, for example, the throne room of Jabba the Hutt (Figure 3.3).

81 From Gwyn Macfarlane, 1979. *Howard Florey: The Making of a Great Scientist.* Oxford University Press; pp. 325–326. Used with permission of Watson, Little Limited, London.

82 Williams, 1984, pp. 146, 150.

83 Drawn from a photograph in Williams, plate 8.

84 See, for example, *The Fly,* in Technocratic Genre Works.

The Last Days of Marcus Karenin

1 The concept of the soft bomb is from Bullard, 1987.

2 Personal communication.

3 The film *Casablanca*, released in 1942, exemplifies the myth described in this section, namely, The Nazi/ the Jew/ and the Yank, because it creates the image of the United States as a Nazi-fighting haven for refugees at a time when few refugees were admitted. However, as it does not specifically reference Jewish refugees, it is not a full-fledged version of the myth. Incidentally, the name *casa blanca*, Spanish for "white house," alludes to American power.

4 Rhodes, p. 197. Also, in Chap. 8, "Stirring and Digging," Rhodes describes Gamow's escape from Moscow.

5 Jonas, pp. 217–222. The figure of $750,000 is on p. 212.

6 See also Lifton's description of the role of the medical profession in the extermination camps themselves: "Medicalized Killing in Auschwitz,"in *The Future of Immortality*, pp. 74-95.

7 See Glass's introduction to H. J. Muller, *Man's Future Birthright,* 1973, p. xi (Technocratic Genre Works).

8 Although born in the United States, Muller, like many of his contemporaries, thought the Bolshevik Revolution was the beginning of the world revolution. He lived and worked in the Soviet Union from 1933 to 1937, until he had a falling-out with Stalin on the issue of Lysenko, and then he left to serve in the Spanish Civil War. Naturally, he lost his American passport during the 1950's. Muller's essay titles, like the episodes in his life, read like an introduction to 20th-century mythology: Evolutionary Advances, Human Freedom, World Peace, Radiation Dangers, Life on Other Planets, and so on.

9 All of this is to be done in a medical context, of course, through genetic counseling. See "What genetic course will man steer?" in the *Proceedings of the Third International Congress of Genetics*, 1966; reprinted in Muller, 1973. The political argument is given on pp. 134 ff.

10 Weaver is quoted by Jonas, *The Circuit Riders,* p. 210. Jonas also describes the founding of the Station for Experimental Evolution at Cold Spring Harbor by the Carnegie Institution in 1904 (p. 121). The Institution's officer in charge of genetics was Charles B. Davenport, an outspoken eugenicist, who served until his retirement in 1934. He also founded the Eugenics Records Office in 1910 to track family histories, with the help of funds from Mrs. E. H. Harriman and the Rockefeller family, which contributed nearly $25,000 to the Eugenics Records Office between 1911 and 1915 (Jonas, p. 134). Harry H. Laughlin, a Davenport associate, testified before Congress in 1920, and this and other expert testimony helped pass the Immigration Restriction Act of 1924 (Jonas, p. 135). But a blue-ribbon committee of appraisal appointed in 1935 subsequently found Davenport's work "scientifically unsatisfactory." Warren Weaver, who was director of the Rockefeller Foundation's scientific patronage from 1932 to 1950 (Jonas, p. 124), emphasized human genetics but eschewed the term eugenics (Jonas, p. 208). The foundation support for Jews began concurrently with Weaver's tenure (Jonas, pp. 213 ff).

11 Barthes, *Mythologies,* "The Brain of Einstein," pp. 68–70. In *The Double Helix*, Watson characterizes Joshua Lederberg as a head that will expand to fill the whole universe.

12 As Dower shows in *War Without Mercy*, the war in the Pacific was waged by both sides with an overtly racist ideology. Whereas in the European theater, Americans discriminated between "Nazis" and "good Germans," in the Pacific theater they maintained that "the only good Jap is a dead Jap." The pervasive racism of Henry Stimson and his associates in the Washington establishment is amply documented by Hodgson (*The Colonel*, pp. 105, 130, 219, 372 ff.); but Stimson's racism, as with that of technocrats generally, was "structural" racism: that is, it forms one of the major premises of institutional behavior but is rarely referred to overtly—except to deny that it is a significant factor in decision-making. For example, Stimson saw World War II as saving "Caucasian civilization" (a curious phrasing) from the Nazis (Hodgson, p. 219), and he helped incarcerate Japanese-Americans in camps; but when pressed on the question of racial prejudice, he claimed to have come from an "Abolitionist family."

13 These statistics are summarized in Wyman, *The Abandonment of the Jews: America and the Holocaust, 1941–1945.*

14 Gross (*Friendly Fascism*) points out that fascism in the United States would be quite different in morphology from its European precursors: it would be "friendly fascism," with winning smiles, hardy handshakes, and computer simulation models—which is another reason why Atwood's book (see below, this chapter) is so misleading. Goethals, in *The Electronic Golden Calf*, identifies these tendencies in the contemporary American mass media.

15 I do not know for a fact that there was only one Jewish admiral, for I have not been able to find any information on this subject. But Wouk, in his novel *The Winds of War*, feels compelled to explain how his hero, a Jewish officer in the United States Navy during World War II, attains a rank high enough to make the book interesting. I do know that the Navy was long considered the personal fiefdom of the American East Coast establishment, and Jewish admirals, while perhaps not impossible, would not have been highly probable either.

16 See Wallace, 1982 (References), and Bacon, 1955 (Technocratic Genre Works).

17 Gray is not Black but not quite White.

18 Mercury Gate is the entrance to the Nevada Test Site where hydrogen bombs are tested. It contrasts to Stallion Gate at the Trinity test site (see the thriller, *Stallion Gate*, by Martin Cruz Smith). The chemical symbol of mercury is Hg.

19 Both Weart (1979, p. 67) and Easlea (1983, p. 150) discuss the role of Wells in the development of 20th-century scientific imagery. See also Easlea (ibid.) on the writings of the now obscure writer Bernal, who historically was perhaps even more important.

20 Wells, 1914, p. 65.

21 Wells, 1914, pp. 97–106.

22 Wells, 1914, p. 106.

23 Wells, 1914, p. 154.

24 Wells, 1914, p. 150.

25 Wells, 1914, p. 156.

26 Wells, 1914, p. 171.
27 Wells, 1914, pp. 174–176. These tables reference the periodic table.
28 Wells, 1914, pp. 185–186.
29 Rhodes, p. 670.
30 Wells, 1914, p. 209.
31 Wells, 1914, p. 248.
32 Frazer assumes, as do all European writers of the period, that "savages" represent an earlier stage of European history, but his emphasis is primarily comparative and mythological.
33 Wells, 1914, p.168.
34 Wells, 1914, p. 264.
35 Wells, 1914, pp. 261–262. The second ellipsis is in the original.
36 Wells, 1914, p. 279. The escape from Russian (or Communist) persecution is another variant of The Nazi/ the Jew/ and the Yank, a myth that Rhodes also chronicles, in the form of Gamow's flight from Moscow.
37 Wells, 1914, p. 299.
38 Significantly, Watson's popularization of his DNA research, *The Double Helix*, is filled with misogynistic references, particularly in regard to Rosalind Franklin, whose research was crucial to his scientific achievements, an issue discussed at length by Sayre (1975). For some general skepticism about the objectivity of scientific medicine, based on comparative study of treatments for diseases in different industrial countries, see Payer (1989). Shorter (1985) has chronicled some disturbing trends towards dehumanization in medicine.
39 Wells, 1914, p. 303.
40 Wells, 1914, pp. 303–304.
41 Wells, 1914, p. 306.
42 Rhodes (p. 676) does not say that Oppenheimer said this when the bomb went off, but he does quote it among the retrospective comments.
43 Quoted by Rhodes, p. 675.
44 Quoted by Rhodes, pp. 571–572. Donne has a minor symbolic role in the technocracy (e.g., Bronowski, p. 406) for reasons that are obscure to me. Carolyn Merchant in *The Death of Nature* discusses Donne's use of mining imagery.
45 Released in 1989. See comments in Technocratic Genre Works.
46 It is significant that Searles (*Films of Science Fiction and Fantasy*, pp. 128–129) questions the orthodoxy of *Close Encounters* because of its "religious overtones." Classic science fiction, as the shadow of scientific atheology, *reduces* gods to intelligent aliens; but in this film Spielberg *assimilates* aliens to gods. Thus, *Close Encounters* marks the transition between science and technocratic eschatology.
47 Laurence's *Men and Atoms*, p. 120. The Christ reference was first published by Laurence in 1946 in *Dawn Over Zero*, perhaps even earlier, but it is downplayed in later works in favor of the second coming of Prometheus.
48 The Ethical Culture background is mentioned in Rhodes, p. 119.
49 The essential idea of the Oppie Effect is that science brings you closer to God—who doesn't exist! A classic embodiment of this doublethink is Capra's *The Tao of Physics: An Exploration of the Parallels Between Modern Physics and Eastern Mysticism* (Technocratic Genre Works), in

which Eastern meditation + Western physics = spirituality. When combined with the myth of the Last Days of Marcus Karenin (see below, this chapter), the Oppie Effect assumes Messianic proportions.

50 Significantly, the major features of the Oppie Effect are adumbrated by Oppie's superior at the War Department, Henry Stimson. As many historians have pointed out, the city of Kyoto was the army's preferred target for the first atomic attack, but Secretary of War Stimson vetoed this choice, not once but twice (Hodgson, pp. 322, 335), on the grounds that Kyoto is the cultural center of Japan. This concern for Kyoto is anomalous in the light of Stimson's decisions to incarcerate Japanese-Americans and to drop the atomic bomb on Japan, decisions which do not indicate any great fondness for the Japanese or their culture; but it is congruent with the theme of "Eastern" spirituality. Also, Stimson makes explicit the relationship between the atomic bomb and Christian salvation. In a conversation with Harvey Bundy (McGeorge's father) in March 1945, Stimson (Hodgson, p. 313) mused on how wonderful it would be if fear of the bomb brought about a revival of Christian principles and if international control were in the hands of someone like Phillips Brooks (1835–1893, Episcopal bishop of Boston).

51 For the comparative anthropology of left/right contrasts in language and mythology, see Rodney Needham's *Right and Left*.

52 Wells, 1914, p. 132.

53 H. G. Wells is a character in the technocratic mythology as well as an author of it. In the film *Time After Time*, released in 1979, Wells (played by Malcolm McDowell) invents the time machine around which the plot revolves (Searles, 1988, pp. 41–43).

54 Symbolic associations to *Carolinum* are important in American nuclear culture. The first proposed site for testing hydrogen bombs within the United States was the outer banks of North Carolina. Later, a plant for the production of tritium was built at Savannah River, South Carolina.

55 Bruce Lincoln, in *Myth, Cosmos, and Society* (1986), shows from surviving texts that the earliest known Indo-European creation myths are characterized by the killing of a mythical hero or totem animal, from whose dismembered body the parts of the world are regenerated. Symbolic anthropology indicates that a similar mythic cycle is part of the iconic deep structure of English too. One possible explanation of these correspondences is that ancient Indo European speakers intellectually elaborated an overt folk model based on an iconic deep structure similar to that of English.

56 The cocoon stage is supported by the remarks that Karenin makes (quoted above) on the night before his surgery. *Cocoon* is also a movie about old people made young again by a visit from friendly aliens.

57 The etymologies in American dictionaries deny or ignore this connection between *testament* and *testicle*.

58 Gribbin, *In Search of the Double Helix*, pp. 190–194.

59 See Cherfas, *Man-made Life*, pp. 6–7, and also Gribben, *In Search of the Double Helix*.

60 Wells, 1923, p. 151. Note that it is cats and women—natural weaklings apparently—who die from exposure to earthlings.

61 In the same vein, Thompson's *Imaginary Landscape* (1989) interprets AIDS as the vehicle which will propel the planet earth to the fourth stage of mathematical consciousness—an enlightened state of mind already attained by Thompson and his friends. Presumably, Thompson's fourth stage of consciousness is equivalent to Wilson's fourth stage of evolution (pure mind), as described in Lumsden and Wilson, *Promethean Fire*. What ever happened to the three stages of everything? See Technocratic Genre Works for bibliographic data on these books.

62 This heading alludes to the much-discussed television series, *Twin Peaks*, created by David Lynch and aired on the ABC network in 1989–1991, which revolves around the unsolved murder of Laura Palmer. The imagery of the show is derived from technocratic mythology (Eastern spirituality, twins, etc.); but as it includes unequivocally magical elements which make no reference to science, such as apparitions by giants and dwarfs, I have not included it in the bibliography of Technocratic Genre Works. However, Lynch did direct the movie *Dune,* a technocratic classic which includes a space-age Messiah.

63 My theory of iconic deep structure postulates that the technocratic hand is shaped differently from a normal human hand, for the pinky is separated from the other fingers by a gap analogous to that of the thumb, which is how I have depicted it.

64 It is said that Oppenheimer's mother also had a withered hand, which she kept hidden by a glove. E. O. Lawrence's mom is famous for being irradiated by her son in the cyclotron.

65 The symbolism of the sexualized middle finger is also found in Greek and Latin (Thass-Thienemann, *Understanding the Symbolic Meaning of Language*, Vol. 1, p. 272).

66 Thass-Thienemann (ibid., pp. 271–274) argues that in Indo-European languages the thumb is symbolically associated with powerful dwarfs who are involved in mining and metallurgy. In English, the thumb is Tom Thumb, a little man.

67 Quoted by Cherfas, 1982, p. 16.

68 These procedures are described in the book by Cherfas, 1982.

69 Rorvik, *In His Image: The Cloning of a Man*,1978 (Technocratic Genre Works). Not just Rorvik but popularizations of molecular biology in general are rife with references to the biblical book of Genesis: titles such as *The Eighth Day of Creation, The Genesis Factor,* and so on.

70 Rorvik's book (1978) is such a mundane rendering of the technocratic mythology and so lacking in the unpredictable details that inevitably give color to historical events that there is little doubt that the book is fiction.

71 Rorvik, 1978, p. 104. Muller won the Nobel prize in 1946 for inducing mutations in fruit flies by means of radiation. Although he first did this work in the 1920's, radiation did not become trendy in biology until after the atomic bomb. For Muller's bioengineering views, see *Man's Future Birthright* (Technocratic Genre Works), as well as Bentley Glass's introduction to this volume. The last essay in Muller's book, first published in 1966 ("What genetic course will man steer?"), is the most pertinent to Rorvik's enthusiasms. Muller was one of James Watson's teachers at the University of Indiana.

72 Rorvik, 1978, p. 32.
73 Rorvik, 1978, p. 80.
74 Rorvik, 1978, p. 181.
75 Rorvik, 1978, p. 201.
76 Rorvik, 1978, p. 205.
77 Stableford, *Future Man*, 1984 (Technocratic Genre Works).
78 Stableford, 1984, p. 12.
79 This sort of special-purpose phenotype is also described in Robin Cook's thriller, *Mutation* (Technocratic Genre Works).
80 Charles DeLisi, "The Human Genome Project," *American Scientist,* 1988 (Technocratic Genre Works).
81 DeLisi, 1988, p. 489.
82 DeLisi, 1988, p. 489.
83 Ball (1986), Fradkin (1989), and Freeman (1982) reproduce declassified government documents, court records, and eyewitness accounts that prove the Atomic Energy Commission (AEC)'s role in the suppression of public health data, the failure to warn people of the dangers of fallout, and the failure to provide medical treatment for people who had been irradiated by nuclear tests. Hilgartner et al. (1983, pp. 87 ff.) show that all Public Health Service press releases on radiation had to be cleared by the AEC, the ancestor of the Department of Energy, which sanitized them to make it appear that there was no health hazard involved in atmospheric testing. These are not obscure historical events but are reported in mainstream American newspapers, as in the articles by Loeb (1990) and Honicker (1989). Significantly, the United States never challenged the validity of these facts in court—but court verdicts awarding damages to victims were nonetheless reversed by higher courts on the ground that citizens cannot sue the government.
84 In Traweek's interviews with particle physicists at SLAC, they complain that bright young people now go into biology.
85 The works by Fletcher (1988), Stark (1990), and Suzuki and Knudson (1990) illustrate how the biotech agenda is implemented as a concern for "ethics" (Technocratic Genre Works).
86 Atwood, *The Handmaid's Tale*, 1986 (References).
87 The movie and the book are different stories. For example, in the book (see References), the heroine is spirited out of the Commander's house by the rebels, with the Commander looking on, whereas in the movie (see Technocratic Genre Works), she stabs the Commander to death before she goes. The movie is clearly technocratic in emphasis (see the annotations), whereas the book is more ambiguous.
88 In *Life Itself*, the father of molecular biology, Francis Crick, argues that all life on earth is extraterrestrial—derived from microbes "seeded" by alien space probes (Technocratic Genre Works).
89 Although some feminists see biotechnology as a progressive force, separating biology from destiny, there are many others who view it, as I do, as the ultimate incorporation of women's bodies into male-dominated institutions and androcentric ideologies. The latter feminist critique has

launched an extensive and hard-hitting literature: Arditti et al. (1984); Birke et al. (1980); Corea (1985); Davis-Floyd (1987, in press); Doane (1990); Rothman (1989); Sault (1990, in press); Spallone (1989); Spallone and Steinberg (1987); and Stolcke (1988).

90 Regis's (1990) bemused style is entertaining, but what he dismisses as "science slightly over the edge" and "fin-de-siècle hubristic mania" is a logical extrapolation of mainstream scientific imagery; and far from being silly, it has potentially serious social and ecological effects. Moreover, it is paid for, more often than not, by taxpayer's money. Cloning and artificial wombs are also conspicuously missing from the list of crackpot schemes.

91 These approaches to immortality are summarized by Regis (1990). See Drexler, *Engines of Creation* (Technocratic Genre Works) for a firsthand account by a major enthusiast.

92 In Stanley Kubrick's movie *2001: A Space Odyssey*, the film ends with the hero transformed into a baby, who floats free in space, as if in a womb. In Spielberg's *Close Encounters of the Third Kind,* the earthlings, bathed in extraterrestrial light, file on to the alien spaceship for a trip to a world beyond time.

93 The steam engine is a spin-off of pure science, for its development presupposed the technique of creating a vacuum through the sudden cooling of steam, an effect which was first created in the laboratory by physicists. See Anthony F. C. Wallace (1982) on the role of government R & D funds in the history of steam power, as well as histories of science for the contributions of Torricelli and Von Guericke to the technology of vacuums.

94 Lifton's title, *The Future of Immortality*, has caught the relationship between the A-bomb and eschatology.

95 For an overview of the cryogenics cult, see the book by Ed Regis (1990). For the religious context of technocratic eschatology in California, see the articles by Lattin (1990).

96 James Watson's book, *The Double Helix* (Technocratic Genre Works) has the latent function of showing that *anyone* can discover the Secret of Life—even Watson. As such, the year of its publication (1968) marks the transition from science disguised as a noble and humanistic calling to a science which admits of no aspirations beyond personal ambition.

97 In the remake of *The Fly,* released in 1986, the scientist-cum-fly attempts to make three persons in one (his girl, their baby, and him) by combining their genetic blueprints.

98 Mojtabai quotes people who hold these beliefs in *Blessèd Assurance*, 1986.

99 *On Human Nature*, pp. 192-193. Note that E. O. Wilson explicitly says that the goal of science is to take over religion's power—not simply to destroy it. Paradoxically, however, Wilson's own brand of biology has no place in the world he envisions, for his own scientific research emphasizes the observation of whole animals in natural habitats, whereas "real" biologists transform pieces of animals using sterile devices. In the pecking order of hard science, sociobiology is contaminated by contact with "nature"—and must be replaced.

The Color Green

1 Personal communication.
2 Gaye Tuchman, *Making News,* 1978.
3 Joseph Needham, in *Science in Traditional China,* shows that gunpowder is a spin-off from Taoist attempts to create an elixir of immortality. Significantly, the first known recipe for gunpowder in European literature, Biringuccio's 16th-century treatise *Pirotechnia* (Technocratic Genre Works), compares it to the Fire of Love.
4 Merchant, *The Death of Nature*, 1983.
5 Wright, *Three Scientists and Their Gods*, pp. 140–141.
6 Yet, there is an optimistic reading too, for when industry has to assure us through expensive prime-time advertising that it is restoring things to the way they were, what better evidence could we have that the myth of development has lost its hold on the populace at large?
7 Barbara Katz Rothman, *Recreating Motherhood*, p. 54.
8 Berman, *The Reenchantment of the World*; Merchant, *The Death of Nature*; and the various works of Easlea.
9 Watson's misogynistic portrait of Rosalind Franklin is well documented by Sayre (1975). Nor have his attitudes toward women softened since 1968, as shown by the choice quotations in Bleier (1986). Many scientists of the older generation objected to Watson's book on the grounds that it portrayed science in an ignoble and unfavorable light, and the Harvard Board of Syndics prevented Harvard University Press from publishing it. Many reviews, both positive and negative, are reproduced in the edition edited by Stent.
10 Significantly, the biologist Richard C. Lewontin, in a review of Watson's book in the Chicago *Sun-Times,* says that the value of the Nobel Prize is based on the myth of the noble scientist. By debunking the myth, Watson debases the coinage of his own life (Stent's edition of *The Double Helix*, p. 187).
11 However, the correspondence to the caduceus is flawed because the serpents twined around Mercury's staff are usually portrayed as opposite in sign—clockwise and anticlockwise—whereas the Double Helix is, with justification, an "SS."
12 Keller (1986) also recognizes that molecular biology, far from discovering the Secret of Life, simply replaced the biology of the prewar period with a vocabulary based on physics and a new set of physicalist questions.
13 For a description of the K/Ar method, see Johanson and Edey (1981, Chap. 9); and Renfrew (1973) for radiometric methods generally. Both books are listed under Technocratic Genre Works because they assimilate anthropology to physical science.
14 See Kamen (1985) for background on the discovery of radiocarbon. Kamen's book is also relevant for other reasons, for the author moved professionally from nuclear physics to molecular biology, and he was one of the victims of the McCarthy period.
15 Quoted in Hilgartner et al., p. 90; see also p.119. Willard F. Libby has published an itemized version of his technocratic agenda, combining both technical and political programs, from which the following sublist was selected:

A. Perhaps he [Man] has come to live in complete harmony with his environment, and to recognize that his true happiness lies in his large control over his environment made possible by his intelligence.
B. Perhaps he has solved the problem of weather control and adjusts the weather to suit his needs, subject only to such natural limitations as the total moisture content of the winds and the total solar flux.
C. Perhaps he has modified his landscape to maximize its usefulness and beauty and has moved mountains to make new river courses to fit his broad design.
D. Perhaps he has solved the problem of aging and lives, on the average, one thousand years, dying only of accidents of one sort or another and retaining his virility and faculties fully to the end.
E. Perhaps atomic energy, both fusion and fission, supplies all of his energy, through large electric-power generating stations and a system of light-weight batteries. Such an intelligent being would have no smog problems, for strict controls would have been put in force to prevent atmospheric pollution with organic matter and the conversion of this matter to harmful products. In such a world, electric cars would be very popular.
F. Perhaps his planet has no deserts, all land being adequately watered either by the controlled rains or by atomic desalting plants on the seacoasts, from which water is pumped inland by atomic power.
G. Perhaps his population is controlled at a comfortable level by matching birth and death rates. The child-bearing period of the women is extended, and birth control is the order of the day, being accomplished with the simplest of mechanical devices.
H. Perhaps the births are handled by petition to the state and permits are granted only to genetically matched parents. Others must use sperm from a sperm bank to assure genetic matching.

From Willard F. Libby, 1965. "Man's place in the physical universe." In *New Views on the Nature of Man*. Edited by John R. Platt. Chicago: University of Chicago Press; pp. 13–14. Copyright © 1965 by The University of Chicago. Used with the permission of the University of Chicago Press.

16 Johannes Fabian, in *Time and the Other*, distinguishes three kinds of time in anthropology. The absolute chronology corresponds to his physical time, pp. 21–22. Incidentally, Fabian's analysis does not presuppose an absolute chronology in the Bible. Rather, the absolute chronology that is replaced by physics is a secular version of sacred history, which many religious commentators believed to be biblically chartered.

17 Not surprisingly, the *international* standard of time, the time of cephalic imagery and high technology, is measured by the frequency of light emitted by glowing atoms (until recently cesium atoms), but *earth* time is chronicled in terms of "half-life."

18 I am aware that quantum physics and relativity theory are said to deny the possibility of measurements that are independent of the position of the observer. However, both of these physical theories assume a culture-free observer who is reduced to his measuring rod—a canonical technocratic image.

REFERENCES

The works in this bibliography illuminate aspects of technocratic cosmology or provide facts that lend themselves to this effort. Some of these works may in fact be technocratic in spirit, but this is not their primary emphasis.

Agricola, Georgius. 1950. *De re metallica: Translated from the first Latin edition of 1556*. Translated by Herbert Clark Hoover and Lou Henry Hoover. New York: Dover Publications.

Aitchison, Leslie. 1960. *A History of Metals.* London: Macdonald and Evans.

Aldred, Cyril. 1984. *The Egyptians.* Revised edition. London: Thames & Hudson.

Alexander. 1988. *Whatever Happened to Divine Grace?* Channeled through Ramón Stevens. Walpole, NH: Stillpoint Publishing.

Arditti, Rita, Renate Duelli Klein, and Shelley Minden, (eds.). 1984. *Test-Tube Women: What Future for Motherhood?* London: Pandora Press.

Arnold, Alan. 1980. *Once Upon a Galaxy: A Journal of the Making of "The Empire Strikes Back."* New York: Ballantine Books.

Atwood, Margaret. 1986. *The Handmaid's Tale.* Boston: Houghton Mifflin.

Ball, Howard. 1986. *Justice Downwind: America's Atomic Testing Program in the 1950's.* Oxford and New York: Oxford University Press.

Barrett, William. 1979. *The Illusion of Technique: A Search for Meaning in a Technological Civilization.* Garden City, NJ: Doubleday Anchor Books. First published in 1978.

Barthes, Roland. 1988. *Mythologies.* Translated by Annette Lavers. New York: The Noonday Press. First published in French in 1957.

Basalla, George. 1989. *The Evolution of Technology.* Cambridge: Cambridge University Press.

Bauschatz, Paul C. 1982. *The Well and the Tree: World and Time in Early Germanic Culture.* N.p.: University of Massachusetts Press.

Bergstein, T. 1972. *Quantum Physics and Ordinary Language.* N.p.: Macmillan.

Berman, Morris. 1981. *The Reenchantment of the World.* New York: Bantam Books.

Birke, Lynda, et al., (eds.) 1980. *Alice Through the Microscope: The Power of Science Over Women's Lives.* London: Virago.

Bleier, Ruth. 1986. Lab coat: robe of innocence or Klansman's sheet? In *Feminist Studies/Critical Studies.* Edited by Teresa de Lauretis. Bloomington, IN: Indiana University Press; pp. 55–66.

Brackman, Arnold C. 1974. *The Dream of Troy*. New York: Van Nostrand Reinhold.

Braudel, Fernand. 1981. *Civilization and Capitalism, 15th to 18th Century*. Vol. 1, *The Structures of Everyday Life*. New York: Harper & Row.

Broad, William J. 1990. "Scientists question history of H-bomb," *San Francisco Chronicle*, January 3, p. A7.

Brown, Robert McAfee. 1983. "The debasement of language," *Christian Century*, April 6: 313–315.

Buchanan, James, and George Watkins. 1976. *The Industrial Archaeology of the Stationary Steam Engine*. London: Allen Lane.

Buchanan, R. A. 1972. *Industrial Archaeology in Britain*. Harmondsworth: Penguin Books.

Bucher, Bernadette. 1981. *Icon and Conquest: A Structural Analysis of the Illustrations of de Bry's "Great Voyages."* Translated by Basia Miller Gulati. Chicago and London: University of Chicago Press.

Bullard, Linda. 1987. Killing us softly: toward a feminist analysis of genetic engineering. In *Made to Order: The Myth of Reproductive and Genetic Progress*. Edited by Patricia Spallone and Deborah Lynn Steinberg. Oxford: Pergamon Press; pp. 110–119.

Cardwell, D. S. L. 1972. *Turning Points in Western Technology*. New York: Neale Watson Academic Publications.

Carnes, Mark C. 1989. *Secret Ritual and Manhood in Victorian America*. New Haven, CT: Yale University Press.

Chomsky, Noam. 1957. *Syntactic Structures*. The Hague: Mouton.

Clark, Grahame. 1977. *World Prehistory in New Perspective*. Third edition. Cambridge: Cambridge University Press.

Clark, Grahame, and Stuart Piggott. 1965. *Prehistoric Societies*. New York: Knopf.

Clifford, James. 1988. *The Predicament of Culture: Twentieth-century Ethnography, Literature, and Art*. Cambridge, MA, and London: Harvard University Press.

Cline, Walter. 1937. *Mining and Metallurgy in Negro Africa*. Menasha, WI: George Banta Co.

Cohn, Carol. 1987. "Nuclear language and how we learned to pat the bomb," *Bulletin of the Atomic Scientists*, June: 17–24.

Coles, John. 1973. *Archaeology by Experiment*. New York: Charles Scribner's Sons.

Congdon, R. J., (ed.) 1977. *Introduction to Appropriate Technology: Toward a Simpler Life-Style*. Emmaus, PA: Rodale Press.

Conklin, Harold C. 1955. "Hanunoó color categories," *Southwestern Journal of Anthropology*, 11, No. 4: 339–344.

Conklin, Harold C. 1972. *Folk Classification: A Topically Arranged Bibliography of Contemporary and Background Reference Through 1971*. New Haven, CT: Department of Anthropology, Yale University.

Connor, Steve, and Sharon Kingman. 1989. *The Search for the Virus: The Scientific Discovery of AIDS and the Quest for a Cure.* Second edition. London: Penguin Books.

Corea, Gena. 1985. *The Mother Machine: Reproductive Technologies from Artificial Insemination to Artificial Wombs.* New York: Harper & Row.

Countryman, L. William. 1988. *Dirt, Greed, and Sex: Sexual Ethics in the New Testament and Their Implications for Today.* Philadelphia: Fortress Press.

Crawford, Harriet, (ed.) 1979. *Subterranean Britain.* New York: St. Martin's Press.

Daniel, Glyn. 1968. *The First Civilizations: The Archaeology of Their Origins.* New York: Thomas Y. Crowell & Co.

Daniel, Glyn. 1968. *The Origins and Growth of Archaeology.* New York: Thomas Y. Crowell & Co.

Davis-Floyd, Robbie E. 1987. "The technological model of birth," *Journal of American Folklore,* October-December: 163–170.

Davis-Floyd, Robbie E. In press. *Birth as an American Rite of Passage.* Berkeley: University of California Press.

de Camp, L. Sprague. 1963. *The Ancient Engineers.* Garden City, NJ: Doubleday.

Delany, Samuel R. 1980. Generic protocols: science fiction and mundane. In *The Technological Imagination: Theories and Fiction.* Edited by Teresa de Lauretis, Andreas Huyssen, and Kathleen Woodward. Madison, WI: Coda Press; pp. 175–193.

Dewey, Joseph. 1990. *In a Dark Time: The Apocalyptic Temper in the American Novel of the Nuclear Age.* West Lafayette, IN: Purdue University Press.

Dickinson, H. W. 1936. *James Watt: Craftsman and Engineer.* Cambridge: Cambridge University Press.

Dickinson, H. W. 1937. *Matthew Boulton.* Cambridge: Cambridge University Press.

Dickinson, H. W. *1939. A Short History of the Steam Engine.* Cambridge: Cambridge University Press.

Diebold, A. Richard, Jr. 1985. "The evolution of Indo-European nomenclature for Salmonid fish: the case of 'huchen' (*Hucho spp.*)," *Journal of Indo-European Studies,* Monograph 5.

Dijkstra, Bram. 1986. *Idols of Perversity: Fantasies of Feminine Evil in Fin-de-Siècle Culture.* New York and Oxford: Oxford University Press.

Doane, Mary Ann. 1990. Technophilia: technology, representation, and the feminine. In *Body/Politics: Women and the Discourses of Science.* Edited by Mary Jacobus, Evelyn Fox Keller, and Sally Shuttleworth. New York and London: Routledge; pp. 163–176.

Douglas, Mary. 1973. *Natural Symbols: Explorations in Cosmology.* Harmondsworth: Penguin. First published in 1970.

Douglas, Mary. 1975. *Implicit Meaning: Essays in Anthropology.* London: Routledge & Kegan Paul.

Dower, John W. 1986. *War Without Mercy: Race and Power in the Pacific War.* New York: Pantheon Books.

Easlea, Brian. 1980. *Witch Hunting, Magic and the New Philosophy: An Introduction to Debates of the Scientific Revolution, 1450–1750.* Atlantic Highlands, NJ: The Humanities Press.

Easlea, Brian. 1983. *Fathering the Unthinkable: Masculinity, Scientists and the Nuclear Arms Race.* London: Pluto Press.

Easlea, Brian. n.d. *Science and Sexual Oppression: Patriarchy's Confrontation With Women and Nature.* London: Weidenfeld and Nicholson.

Eco, Umberto. 1989. *Foucault's Pendulum.* Translated by William Weaver. New York: Harcourt Brace Jovanovich. First published in Italian in 1988.

Eiseley, Loren. 1961. *Darwin's Century: Evolution and the Men Who Discovered It.* Garden City, NJ: Doubleday Anchor Books.

Ellul, Jacques. 1964. *The Technological Society.* Translated by John Wilkinson. New York: Vintage Books. First published in French in 1954.

Fabian, Johannes. 1983. *Time and the Other: How Anthropology Makes Its Object.* New York: Columbia University Press.

Fagan, Brian. 1978. *Elusive Treasure: The Story of Early Archaeologists in the Americas.* N.p.: Book Club Associates.

Fagan, Brian M. 1975. *The Rape of the Nile: Tomb Robbers, Tourists, and Archaeologists in Egypt.* New York: Simon & Schuster.

Fanon, Frantz. 1970. *Black Skin White Masks.* Translated by Charles Lam Markmann. N.p.: Paladin Books.

Feyerabend, Paul. 1987. *Farewell to Reason.* London and New York: Verso.

Finn, R. Welldon. 1963. *An Introduction to the Domesday Book.* Bishop Briggs: Villafield Press.

Firth, Raymond. 1965. *Primitive Polynesian Economy.* Second edition. London: Routledge & Kegan Paul. First published in 1939.

Firth, Raymond. 1967. *The Work of the Gods in Tikopia.* Second edition. New York: Humanities Press.

Flood, Josephine. 1983. *Archaeology of the Dream Time.* Honolulu: University of Hawaii Press.

Florman, Samuel C. 1976. *The Existential Pleasures of Engineering.* New York: St. Martin's Press.

Forbes, R. J. 1955. *Studies in Ancient Technology.* Leiden: E. J. Brill.

Forbes, R. J. 1958. *Man the Maker: A History of Technology and Engineering.* London and New York: Abelard-Sherman Ltd.

Forde, C. Daryll. 1963. *Habitat, Economy and Society: A Geographical Introduction to Ethnology.* New York: E. P. Dutton. First published in 1934.

Fradkin, Philip L. 1989. *Fallout: An American Nuclear Tragedy.* N.p.: University of Arizona Press.

Frake, Charles O. 1962. The ethnographic study of cognitive systems, with discussion by Harold C. Conklin. In *Anthropology and Human Behavior*. Edited by Thomas Gladwin and William C. Sturtevant. Washington, D. C.: The Anthropological Society of Washington; pp. 72–93.

Freeman, Leslie J. 1982. *Nuclear Witnesses: Insiders Speak Out*. New York and London: W. W. Norton & Co.

French, James Kip. 1960. *The Story of Engineering*. Garden City, NJ: Doubleday.

Freud, Sigmund. 1975. *The Psychopathology of Everyday Life*. Translated by Alan Tyson. Harmondsworth: Penguin Books. First published in German in 1901.

Freud, Sigmund. 1976. *The Interpretation of Dreams*. Translated by James Strachey. Harmondsworth: Penguin Books. First published in German in 1900.

Freud, Sigmund. 1976. *Jokes and Their Relation to the Unconscious*. Translated by James Strachey. Harmondsworth: Penguin Books. First published in German in 1905.

Funkenstein, Amos. 1986. *Theology and the Scientific Imagination from the Middle Ages to the Seventeenth Century*. Princeton, NJ: Princeton University Press.

Galileo. 1989. *Sidereus Nuncius or The Sidereal Messenger*. Translated by Albert Van Helden. Chicago and London: University of Chicago Press. A translation of the Venice edition of 1610.

Gillispie, Charles C., (ed.) 1959. *A Diderot Pictorial Encyclopedia of Trades and Industry*, 2 volumes. New York: Dover Publications.

Gillispie, Charles C. 1959. *Genesis and Geology: A Study of the Relations of Scientific Thought, Natural Theology, and Social Opinion in Great Britain, 1790–1850*. New York: Harper and Row. First published in 1951.

Gillispie, Charles C., (ed.) 1970–1980. *Dictionary of Scientific Biography*. New York: Charles Scribner's Sons.

Gimbutas, Marija. 1965. *Bronze Age Cultures in Central and Eastern Europe*. The Hague: Mouton.

Gimpel, Jean. 1976. *The Medieval Machine*. Harmondsworth: Penguin Books. First published in French in 1974.

Glob, P. V. 1969. *The Bog People: Iron-Age Man Preserved*. London: Faber & Faber. First published in Danish in 1965.

Glob, P. V. 1974. *The Mound People: Danish Bronze-Age Man Preserved*. London: Faber & Faber. First published in Danish in 1970.

Goethals, Gregor T. 1990. *The Electronic Golden Calf: Images, Religion, and the Making of Meaning*. Cambridge, MA: Cowley Publications.

Grant, Michael. 1971. *The Cities of Vesuvius: Pompeii and Herculaneum*. London: Weidenfeld & Nicholson.

Griffin, Susan. 1981. *Pornography and Silence: Culture's Revenge Against Nature*. New York: Harper & Row.

Gross, Bertram. 1980. *Friendly Fascism: The New Face of Power in America.* Boston: South End Press.

Gyi, M. 1984. "Semantics of nuclear politics," *Et Cetera,* Summer: 135–147.

Hall, Marie Boas. 1971. *The Pneumatics of Hero of Alexandria: A Facsimile of the 1851 Edition.* New York: American Elsevier.

Hare, Ronald. 1970. *The Birth of Penicillin and the Disarming of Microbes.* London: George Allen and Unwin Ltd.

Hendricks, Rhoda H. 1974. *Classical Gods and Heroes: Myths as Told by the Ancient Authors.* New York: Morrow Quill Paperbacks.

Heriot, M. Jean. 1989. *Blessed Assurance: Assessing Religious Beliefs through Actions in a Carolina Baptist Church.* University of California, Los Angeles. Distributed by UMI, Ann Arbor, MI.

Hilgartner, Stephen, Richard C. Bell, and Rory O'Connor. 1983. *Nukespeak: The Selling of Nuclear Technology in America.* Harmondsworth: Penguin. First published in 1982.

Hodges, Henry. 1974. *Technology in the Ancient World.* New York: Knopf.

Hodgson, Godfrey. 1990. *The Colonel: The Life and Wars of Henry Stimson, 1867–1950.* New York: Alfred A. Knopf.

Honicker, Clifford T. 1989. "America's radiation victims: the hidden files," *The New York Times Magazine,* November 10: 38–41, 98–103, 120.

Hunter, Louis C. 1979. *Waterpower: A History of Industrial Power in the United States.* Vol. 3, *1780–1830.* Charlottesville, VA: University of Virginia Press.

Jakobson, Roman, and M. Halle. 1956. *Fundamentals of Language.* The Hague: Mouton.

Jonas, Gerald. 1989. *The Circuit Riders: Rockefeller Money and the Rise of Modern Science.* New York and London: W.W. Norton and Company.

Kamen, Martin D. 1985. *Radiant Science, Dark Politics: A Memoir of the Nuclear Age.* Berkeley and Los Angeles: University of California Press.

Keegan, John. 1976. *The Face of Battle.* New York: Viking Press.

Keen, Benjamin. 1971. *The Aztec Image in Western Thought.* New Brunswick, NJ: Rutgers University Press.

Keller, Evelyn Fox. 1986. Making gender visible in the pursuit of nature's secrets. In *Feminist Studies/Critical Studies.* Edited by Teresa de Lauretis. Bloomington, IN: Indiana University Press; pp. 67–77.

Keuls, Eva C. 1985. *The Reign of the Phallus: Sexual Politics in Ancient Athens.* New York: Harper & Row.

Kline, Morris. 1980. *Mathematics: The Loss of Certainty.* New York: Oxford University Press.

Kroeber, Theodora. 1967. *Ishi in Two Worlds.* Berkeley and Los Angeles: University of California Press.

Kuhn, Thomas S. 1970. *The Structure of Scientific Revolutions.* Chicago: University of Chicago Press.

Lakoff, George. 1987. *Women, Fire, and Dangerous Things: What Categories Reveal about the Mind.* Chicago: University of Chicago Press.

Larimer, Tim. 1990. "The next ice age may be starting here in Silicon Valley as techies turn to cryonics so they can download their intellects in the distant future," *San Jose Mercury News,* December 9, *West* magazine, pp. 17–26.

Lattin, Don. 1990. "New Age mysticism strong in Bay Area: just 44% hold traditional view of God," *San Francisco Chronicle,* April 24, A1, A8.

Lattin, Don. 1990. "Warning of New Age 'Threat'," *San Francisco Chronicle,* April 25, A1, A6.

Leach, Edmund. 1969. *Genesis as Myth and Other Essays.* London: Jonathan Cape.

Leach, Edmund. 1970. *Lévi-Strauss.* N.p.: Fontana.

Leach, Edmund. 1973. Levels of communication and problems of taboo in the appreciation of primitive art. In *Primitive Art and Society.* Edited by Anthony Forge. Oxford: Oxford University Press; pp. 231–234.

Leach, Edmund, and D. Alan Aycock. 1983. *Structuralist Interpretations of Biblical Myth.* Cambridge: Cambridge University Press.

Leach, Edmund R. 1964. Anthropological aspects of language: animal categories and verbal abuse. In *New Directions in the Study of Language.* Edited by Eric H. Lenneberg. Cambridge, MA: M.I.T. Press; pp. 23–63.

Lévi-Strauss, Claude. 1962. *The Savage Mind.* Chicago: University of Chicago Press. First published in French in 1962.

Lévi-Strauss, Claude. 1967. *Structural Anthropology.* Translated by Claire Jacobson and Brooke Grundfest Schoepf. Garden City, NY: Doubleday. First published in French in 1963.

Lévi-Strauss, Claude. 1969. *The Raw and the Cooked: Introduction to a Science of Mythology,* Vol. 1. Translated by John and Doreen Weightman. Chicago: University of Chicago Press. First published in French in 1964.

Lévi-Strauss, Claude. 1973. *From Honey to Ashes: Introduction to a Science of Mythology,* Vol. 2. Translated by John and Doreen Weightman. Chicago: University of Chicago Press. First published in French in 1966.

Lewis, Arthur O., Jr. 1963. *Of Men and Machines.* New York: E. P. Dutton.

Lifton, Robert Jay. 1987. *The Future of Immortality and Other Essays for a Nuclear Age.* New York: Basic Books.

Lincoln, Bruce. 1986. *Myth, Cosmos, and Society: Indo-European Themes of Creation and Destruction.* Cambridge, MA: Harvard University Press.

Lloyd, G. E. R. 1979. *Magic, Reason and Experience: Studies in the Origin and Development of Greek Science.* Cambridge: Cambridge University Press.

Loeb, Paul. 1990. "From Hanford, People Learned the Value of Life," San Jose *Mercury News,* August 26, 1C, 8C.

Lovejoy, Arthur O. 1960. *The Great Chain of Being: A Study of the History of an Idea.* New York: Harper Torchbooks. First published in 1936.

Lovell, Bernard. 1981. *Emerging Cosmology*. New York: Columbia University Press.

Macfarlane, Gwyn. 1979. *Howard Florey: The Making of a Great Scientist*. Oxford: Oxford University Press.

MacKendrick, Paul. 1960. *The Mute Stones Speak: The Story of Archaeology in Italy*. New York: St. Martin's Press.

Malinowski, Bronislaw. 1961. *Argonauts of the Western Pacific*. New York: E. P. Dutton. First published in 1922.

Marsden, E. W. 1969. *Greek and Roman Artillery: Historical Development*. London: Oxford University Press.

Marshak, Alexander. 1972. *The Roots of Civilization: The Cognitive Beginnings of Man's First Art, Symbols, and Notation*. New York: McGraw-Hill.

Martin, Emily. 1987. *The Woman in the Body: A Cultural Analysis of Reproduction*. Boston: Beacon Press.

Martin, Emily. 1990. Science and women's bodies: forms of anthropological knowledge. In *Body/Politics: Women and the Discourses of Science*. Edited by Mary Jacobus, Evelyn Fox Keller, and Sally Shuttleworth. London and New York: Routledge; pp. 69–82.

Mayer, Arno J. 1988. *Why Did the Heavens Not Darken?* New York: Pantheon.

McCarty, Maclyn. 1985. *The Transforming Principle: Discovering that Genes Are Made of DNA*. New York: W. W. Norton.

McNaughton, Patrick R. 1988. *The Mande Blacksmiths: Knowledge, Power, and Art in West Africa*. Bloomington and Indianapolis: Indiana University Press.

Mellaart, James. 1965. *Earliest Civilizations of the Near East*. New York: McGraw-Hill.

Merchant, Carolyn. 1983. *The Death of Nature: Women, Ecology, and the Scientific Revolution*. New York: Harper & Row.

Meyer, Alfred. 1980. "Temple of the Aztecs," *Science 80*, November: 67–73.

Millard, Ann V. 1990. "The place of the clock in pediatric advice: rationales, cultural themes, and impediments to breastfeeding," *Journal of Social Science and Medicine*, 31, No. 2: 211–221.

Mojtabai, A. G. 1986. *Blessèd Assurance: At Home With the Bomb in Amarillo, Texas*. Boston: Houghton Mifflin Co.

Montet, Pierre. 1958. *Everyday Life in Egypt*. London: St. Martin's Press.

Morgenthaler, Eric. 1990. "For a healthy glow, some old folks try a dose of radon: Montana mine set swears by the radioactive 'cure'...," *The Wall Street Journal*, October 12, A1, A8.

Mumford, Lewis. 1934. *Technics and Civilization*. New York: Harcourt, Brace & World.

Munn, Nancy. 1973. *Walbiri Iconography*. Ithaca, NY: Cornell University Press.

Nash, Hugh, (ed.) 1977. *Progress As If Survival Mattered: A Handbook for a Conserver Society*. San Francisco: Friends of the Earth.

Nebelsick, Harold P. 1985. *Circles of God: Theology and Science from the Greeks to Copernicus.* Edinburgh: Scottish Academic Press.

Needham, Joseph. 1961. *Classical Chinese Contributions to Mechanical Engineering.* 41st Earl Gray Memorial Lecture.

Needham, Joseph. 1964. *The Development of Iron and Steel Technology in China.* N.p.: The Newcomen Society and W. Heffer & Sons.

Needham, Joseph. 1981. *Science in Traditional China: A Comparative Perspective.* Cambridge, MA: Harvard University Press.

Needham, Rodney. 1967. Blood, thunder, and the mockery of animals. In *Myth and Cosmos: Readings in Mythology and Symbolism.* Edited by John Middleton. Garden City, NJ: The Natural History Press; pp. 271–285.

Needham, Rodney, (ed.) 1973. *Right and Left: Essays on Dual Symbolic Classification.* Chicago and London: University of Chicago Press.

Orwell, George. 1949. *1984.* New York: Harcourt Brace Jovanovich.

Orwell, George. 1968. Politics and the English language. In *The Collected Essays, Journalism and Letters of George Orwell,* Vol. 4. Edited by Sonia Orwell and Ian Angus. New York and London: Harcourt Brace Jovanovich; pp. 127–140. First published in 1946.

Parker, S. Thomas. 1984. "Exploring the Roman frontier in Jordan," *Archaeology,* Sept/Oct: 33–39.

Paul, Charles B. 1980. *Science and Immortality: The Éloges of the Paris Academy of Sciences (1699–1791).* Berkeley and Los Angeles: University of California Press.

Payer, Lynn. 1989. *Medicine and Culture: Varieties of Treatment in the United States, England, West Germany, and France.* New York: Penguin Books. First published in 1988.

Polkinghorne, John. 1989. *Science and Creation: The Search for Understanding.* Boston: Shambala.

Polkinghorne, John. 1989. *Science and Providence: God's Interaction with the World.* Boston: Shambala.

Pope, Saxton T. 1962. *Bows and Arrows.* Berkeley and Los Angeles: University of California Press. First published in 1923.

Preston, Richard. 1987. "Reporter at large: dark time," *New Yorker,* October 26: 64–97.

Pyles, Thomas. 1971. *The Origins and Development of the English Language.* Second edition. New York: Harcourt Brace Jovanovich.

Redondi, Petro. 1987. *Galileo Heretic.* Translated by Raymond Rosenthal. Princeton, New Jersey: Princeton University Press. First published in Italian in 1983.

Regis, Ed. 1987. *Who Got Einstein's Office? Eccentricity and Genius at the Institute for Advanced Study.* Reading, MA: Addison-Wesley Publishing Company.

Regis, Ed. 1990. *Great Mambo Chicken and the Transhuman Condition: Science Slightly Over the Edge.* Reading, MA: Addison-Wesley Publishing Company.

Reynolds, Peter C. 1981. *On the Evolution of Human Behavior: The Argument from Animals to Man.* Berkeley and Los Angeles: University of California Press.

Reynolds, Peter C. 1991. Structural differences in intentional action between humans and chimpanzees: with their implications for theories of handedness and bipedalism. In *Semiotic Modeling.* Edited by Myrdene Anderson and Floyd Merrell. Berlin: Walter de Gruyter & Co.

Rifkin, Jeremy. 1983. *Algeny.* New York: Viking Press.

Robbins, Gary. 1990. "Tumor victim sues to have his head preserved: he hopes freezing will give science time to find a cure," Orange County Register News Service.

Robinson, Eric, and A. E. Musson. 1969. *James Watt and the Steam Revolution.* London: Adams & Dart.

Rolt, L. T. C. 1963. *Thomas Newcomen: The Prehistory of the Steam Engine.* London: Macdonald.

Rolt, L. T. C., and J. S. Allen. 1977. *The Steam Engine of Thomas Newcomen.* New York: Science History Publications.

Romanyshyn, Robert D. 1989. *Technology as Symptom and Dream.* London and New York: Routledge.

Rosenthal, Michael. 1986. *The Character Factory: Baden-Powell's Boy Scouts and the Imperatives of Empire.* New York: Pantheon Books.

Rothman, Barbara Katz. 1989. *Recreating Motherhood: Ideology and Technology in a Patriarchal Society.* New York and London: W. W. Norton & Co.

Rushforth, Keith. 1981. *The Pocket Guide to Trees.* New York: Simon and Schuster.

Said, Edward W. 1979. *Orientalism.* New York: Vintage Books. First published in 1978.

Salisbury, R. F. 1962. *From Stone to Steel: Economic Consequences of Technological Change in New Guinea.* London: Cambridge University Press.

Sandars, N. K. 1985. *The Sea Peoples: Warriors of the Ancient Mediterranean.* Revised edition. London: Thames & Hudson.

Sandmel, Samuel, M. Jack Suggs, and Arnold J. Tracik, (eds.) 1976. *The New English Bible with the Apocrypha: Oxford Study Edition.* New York: Oxford University Press.

Sault, Nicole. 1990. "Walking wombs: surrogate motherhood as political control," Presented at the Annual Meeting of the American Anthropological Association, New Orleans.

Sault, Nicole. In press. Surrogate mothers and godmothers: cultural definitions of parenthood and the body in the United States and Mexico. In *Many Mirrors: Body Image and Social Relations in Anthropological Perspective.* Edited by Nicole Sault. Philadelphia: University of Pennsylvania Press.

Sayre, Anne. 1975. *Rosalind Franklin and DNA.* New York: W. W. Norton.

Searles, Baird. 1988. *Films of Science Fiction and Fantasy*. New York: Harry N. Abrams.

Shilts, Randy. 1988. *And the Band Played On: Politics, People, and the AIDS Epidemic*. New York: Penguin Books.

Shorter, Edward. 1985. *Bedside Manners: The Troubled History of Doctors and Patients*. New York: Simon & Schuster.

Soedel, Werner, and Vernard Foley. 1979. "Ancient catapults," *Scientific American*, March: 120–128.

Sontag, Susan. 1990. *Illness as Metaphor and AIDS and Its Metaphors*. New York: Doubleday Anchor Books. First published in 1978 and 1989 respectively.

Spallone, Patricia. 1989. *Beyond Conception: The New Politics of Reproduction*. London: Macmillan.

Spallone, Patricia, and Deborah Lynn Steinberg, (eds.) 1987. *Made to Order: The Myth of Reproductive and Genetic Progress*. Oxford: Pergamon Press.

Stevens, Ramón. 1988. *Whatever Happened to Divine Grace?* Walpole, NH: Stillpoint Publishing.

Stolcke, Verena. 1988. "New reproductive technologies: the old quest for fatherhood," *Reproductive and Genetic Engineering*, 1, No. 1: 5–19.

Strathern, Andrew. 1979. *Ongka: A Self-Account by a New Guinea Big-Man*. New York: St. Martin's Press.

Taton, René, (ed.) 1963. *History of Science: Ancient and Medieval Science From the Beginnings to 1450*. Translated by A. J. Pomerans. New York: Basic Books. First published in French in 1957.

Thass-Thienemann, Theodore. 1973. *The Interpretation of Language*. Vol. 1, *Understanding the Symbolic Meaning of Language*. New York: Jason Aronson, Inc.

Thass-Thienemann, Theodore. 1983. *Understanding the Unconscious Meaning of Language*. New York: Jason Aronson.

Theweleit, Klaus. 1989. *Male Fantasies*. Translated by Erica Carter and Chris Turner. Minneapolis: University of Minnesota Press. First published in German in 1978.

Thomas, Gordon, and Max Morgan Witts. 1977. *Enola Gay*. New York: Stein and Day.

Toulmin, Stephen. 1985. *The Return to Cosmology: Postmodern Science and the Theology of Nature*. Berkeley and Los Angeles: University of California Press.

Traweek, Sharon. 1988. *Beamtimes and Lifetimes: The World of High Energy Physicists*. Cambridge, MA, and London: Harvard University Press.

Triewald, Marten. 1928. *A Short Description of the Atmospheric Engine*. Cambridge: W. Heffer & Sons. First published in Swedish in 1734.

Tuchman, Gaye. 1978. *Making News: A Study in the Construction of Reality*. New York: The Free Press.

Van der Merwe, Nikolaas, and Donald J. Avery. 1982. "Pathways to Steel," *American Scientist*, 70, March/April: 146–155.

Wallace, Anthony F. C. 1982. *The Social Context of Innovation: Bureaucrats, Families, and Heroes in the Early Industrial Revolution, as Foreseen by Bacon's "New Atlantis."* Princeton: Princeton University Press.

Wallerstein, Immanuuel. 1974. *The Modern World-System: Capitalist Agriculture and the Origins of the European World-Economy in the Sixteenth Century*. London: Academic Press.

Ward-Perkins, J. B. 1972. "Quarrying in antiquity: technology, tradition and social change," *Proceedings of the British Academy*, 62, Mortimer Wheeler Lecture, 1971.

Watkins, Calvert, (ed.) 1985. *The American Heritage Dictionary of Indo-European Roots*. Boston: Houghton Mifflin.

Weart, Spencer. 1979. *Scientists in Power*. Cambridge, MA, and London: Harvard University Press.

Weart, Spencer R. 1988. *Nuclear Fear: A History of Images*. Cambridge, MA, and London: Harvard University Press.

Weaver, Warren. 1970. *Scene of Change: A Lifetime in American Science*. New York: Charles Scribner's Sons.

Webster, Graham. 1969. *The Roman Imperial Army of the First and Second Centuries A.D.* London: Adam and Charles Black.

Wendland, Albert. 1985. *Science, Myth, and the Fictional Creation of Alien Worlds*. Ann Arbor, MI: UMI Research Press.

Wertime, Theodore A., and James D. Muhly, (eds.) 1980. *The Coming of the Age of Iron*. New Haven, CT: Yale University Press.

White, J. Peter, and Jim Allen. 1980. "Melanesian Prehistory: Some Recent Advances," *Science*, 207, 728–734.

White, Peter, and James O'Connell. 1984. *Prehistory of Australia, New Guinea, and the Sahul*. New York: Academic Press.

Williams, Joseph M. 1975. *Origins of the English Language: A Social and Linguistic History*. New York: The Free Press and Macmillan Publishing Co.

Williams, Trevor I. 1984. *Howard Florey: Penicillin and After*. Oxford: Oxford University Press.

Wilson, John A. 1951. *The Culture of Ancient Egypt*. Chicago: University of Chicago Press.

Winner, Langdon. 1977. *Autonomous Technology: Technics-Out-of-Control as Theme in Political Thought*. Cambridge, MA: M.I.T. Press.

Wyman, David S. 1984. *The Abandonment of the Jews: America and the Holocaust, 1941–1945*. New York: Pantheon Books.

Yates, Frances A. 1964. *Giordano Bruno and the Hermetic Tradition*. Chicago: University of Chicago Press.

Young, Michael W. 1971. *Fighting With Food: Leadership, Values, and Social Control in a Massim Society*. Cambridge: Cambridge University Press.

TECHNOCRATIC GENRE WORKS

These books, articles, videos, and films exemplify the technocratic mythology. Technocratic myth is not restricted to specific literary genres but constitutes a genre of its own. This bibliography includes works of fiction and nonfiction, works of science and sci-fi, history and journalism, academic treatises as well as popularizations. As the technocratic mythology is not a recent development but a cultural tradition of considerable antiquity, this bibliography spans a period of three hundred years, including such works as Biringuccio's 16th-century treatise on gunpowder, La Mettrie's 18th-century discourse on the human machine, various 19th-century theories of human evolution, some 20th-century science fiction, and recent scientific popularizations. This bibliography is intended to be representative, not complete, and many more examples could be found, especially in technocratic house organs such as Scientific American *and* Omni *magazine, as well as in the technical publications of mainstream science itself.*

1942. *Casablanca.* Directed by Michael Curtiz. Warner Brothers.

This film straddles the boundary of technocratic mythology: An archetypal American (Humphrey Bogart) confronts the Nazis, gives up love (Ingrid Bergman), and goes off to Brazzaville (in the Congo) to fight for freedom. Also, it presents the United States as a haven for refugees at a time when the borders were essentially closed.

1958. *The Fly.* A film produced and directed by Kurt Neumann. Twentieth Century-Fox.

In this 1950's sci-fi classic, a human and a fly accidentally get their heads swapped in a scientist's time machine.

1971. *The Andromeda Strain.* Directed by Robert Wise. Based on the novel by Michael Crichton. Universal Pictures.

A sci-fi film about an alien virus that can only be safely contained using a germ warfare laboratory. The scientific rituals of purity and pollution are highly visualized.

1974. *Terminal Man.* A film produced and directed by Mike Hodges. Based on the novel by Michael Crichton. Warner Brothers.

The surgery scene epitomizes scientific medicine and the human body as machine.

1977. *Close Encounters of the Third Kind.* Written and directed by Steven Spielberg. Columbia Pictures.

An alien spaceship, awash in light and mandalas, appears to a select group of technocrats and believers at a secret American base at Devil's Tower,

Wyoming. A squadron of American flyers who had disappeared in 1945 emerge from the ship—unharmed and not a day older.

1977. *Star Wars: Episode IV: A New Beginning.* A film directed by George Lucas. Produced by Twentieth Century-Fox.

This film and its sequels are the definitive presentation of American technocratic imagery.

1980. *Star Wars: The Empire Strikes Back.* A film with a story by George Lucas. Lucasfilm and Twentieth Century-Fox.

1983. *Star Wars: Return of the Jedi.* A film with a story by George Lucas. Lucasfilm and Twentieth Century-Fox.

1984. *Dune.* Directed by David Lynch. Screenplay by David Lynch. Based on the novel by Frank Herbert. Universal Pictures.

In this stately sci-fi epic, a priestess (with "control of bloodlines") predicts a future messiah who will lead the oppressed peoples of the desert planet (Arakis or Dune) to "true freedom." After the putative messiah is proclaimed the Hand of God, he launches a quick and successful atomic attack on the evil empire—gaining control of a mysterious elixir of life and flooding the desert with water.

1986. *The Fly.* A film directed by David Cronenberg. A Brooksfilm Production. Twentieth Century-Fox.

In this remake of the '50's film, the scientist-cum-fly tries to genetically fuse himself, his girlfriend, and their unborn child into a single body—a man-made trinity.

1989. *The Handmaid's Tale.* A film directed by Volker Schlondorf. Screenplay by Harold Pinter. Based on the novel by Margaret Atwood. Cinecom.

This film, a technocratic transformation of Atwood's novel, from which it differs in significant ways, portrays surrogate motherhood as a consequence of Christianity, presents religion as the source of totalitarianism, and equates the absence of sexual repression with political liberty.

1989. *Indiana Jones and the Last Crusade.* A film produced by Steven Spielberg. Paramount-Lucasfilm.

This film presents some of the key images of the Manhattan Project in a covert, fictionalized form.

Anonymous. 1989. "The future and you: a 30-page preview of 2000 and beyond," *Life*, February: passim.

This is hard-core technoporn. The cover art shows a male head split down the middle, with one side natural, the other an android. The pictures inside are even more revealing.

Ansley, David, and Alex Barnum. 1990. "Brave new biotech," San Jose *Mercury News,* December 30, *West* magazine, pp. 6–11.

This article lists the "top 10 coming attractions" in biotechnology—environmentalist nightmares presented as examples of technical progress.

Bacon, Francis. 1955. *Selected Writings of Francis Bacon, with an Introduction and Notes by Hugh G. Dick.* New York: Random House.

This 17th-century author created the modern version of Nebuchadnezzar's Dream.

Barnaby, Frank. 1971. *Man and the Atom: The Uses of Nuclear Energy*. New York: Funk and Wagnalls.

This popularization shows how the pure research of nuclear physics is to be converted into artifacts that exemplify Nebuchadnezzar's Dream.

Biringuccio, Vannoccio. 1942. *The Pirotechnia*. Translated by Cyril Stanley Smith and Martha Teach Gnudi. New York: The American Institute of Mining and Metallurgical Engineers.

This 16th-century technologist romanticizes gunpowder as the fire that burns without leaving ashes.

Bronowski, J. 1973. *The Ascent of Man*. Boston and Toronto: Little, Brown and Company.

This spin-off book from the BBC television series is a highly polished presentation of the myth of The Lone Galileo.

Burke, James. 1978. *Connections*. Boston and Toronto: Little, Brown and Company.

Although this spin-off book from the BBC television series tries hard to replicate Bronowski's *The Ascent of Man* (compare the back flaps of the dust jackets), it simply confirms the transformation of science into mindless technical change—in less than a decade.

Campbell, Joseph. 1951. *The Flight of the Wild Gander: Explorations of the Mythical Dimension*. South Bend, IN: Regnery/Gateway.

This book makes explicit the technocratic premises of Campbell's work: progressive stages of mythic development, the triumph of secularization, the atheological agenda, etc.

Campbell, Joseph. 1990. *Transformations of Myth Through Time*. New York: Harper & Row.

A spin-off from the PBS television series, this book reinforces the technocratic image of myth as remote from science.

Capra, Fritjov. 1975. *The Tao of Physics: An Exploration of the Parallels Between Modern Physics and Eastern Mysticism*. Berkeley: Shambhala.

Eastern mysticism + Western physics = spirituality.

Casti, John L. 1989. *Paradigms Lost: Images of Man in the Mirror of Science*. New York: William Morrow and Company.

This is a good example of the technocratic popularization. Important philosophical and theological questions, such as the origin of life or the nature of mind, are explicated strictly within the confines of what scientists have to say about them.

Cherfas, Jeremy. 1982. *Man-made Life: An Overview of the Science, Technology and Commerce of Genetic Engineering*. New York: Pantheon Books.

An objective presentation of neutral science—with all that such a characterization implies.

Childe, V. Gordon. 1951. *Social Evolution*. London: Watts.

Human evolution interpreted as Nebuchadnezzar's Dream.

Childe, V. Gordon. 1957. *The Dawn of European Civilization*. Sixth edition. New York: Vintage Books.

The use of the stage model in archaeology.

Cook, Robin. 1989. *Mutation.* New York: G. P. Putnam's Sons.
A thriller about a secret project to clone humans.

Crichton, Michael. 1980. *Congo.* New York: Avon Books.
A sci-fi thriller that highlights the symbolic role of the Congo in technocratic mythology. The racism is overt, as are the references to the Manhattan Project.

Crichton, Michael. 1987. *Sphere.* New York: Ballantine Books.
The Y-forked organ in marine aspect: a huge cylinder containing a mysterious sphere is found embedded in the sea floor.

Crick, Francis. 1981. *Life Itself: Its Origin and Nature.* New York: Simon and Schuster.
The father of molecular biology denies the existence of terrestrial life: "life here was seeded by microorganisms sent on some form of spaceship by an advanced civilization" (p. 141).

Daley, Brian. 1979. *Han Solo at Star's End: From the Adventures of Luke Skywalker.* New York: Ballantine Books.
A spin-off book from the *Star Wars* film.

Daley, Brian. 1980. *Han Solo and the Lost Legacy: From the Adventures of Luke Skywalker.* New York: Ballantine.
A spin-off book from the *Star Wars* film.

DeLisi, Charles. 1988. "The Human Genome Project," *American Scientist,* 76, No. 5, September-October: 488-493.
A charter account of HGI for a scientific audience.

Drexler, K. Eric. 1986. *Engines of Creation.* Garden City, NY: Anchor Press/Doubleday.
Tiny, molecular "nanomachines" will replace the biological cell—abolishing disease, conquering death, and colonizing outer space. This will increase Freedom for all mankind.

Einstein, Albert. N.d. *The World As I See It.* Translated by Alan Harris. Secaucus, NJ: Citadel Press.
The brain at work: Einstein as guru.

Engels, Friedrich. 1942. *The Origin of the Family, Private Property, and the State: In the Light of the Researches of Lewis H. Morgan.* New York: International Publishers.
First published in 1884, this work assimilates Morgan's version of Nebuchadnezzar's Dream to a Marxist version of Nebuchadnezzar's Dream.

Firestone, Shulamith. 1971. *The Dialectic of Sex: The Case for Feminist Revolution.* New York: Bantam Books.
First published in 1970, Firestone's book assimilates feminist goals to a stage-successive cosmogony. See the chart of society and history on the last page of the book.

Flanagan, Dennis. 1988. *Flanagan's Version: A Spectator's Guide to Science on the Eve of the 21st Century.* New York: Alfred A. Knopf.
This book by a co-founder of *Scientific American* has many examples of that indomitable denizen of the popularization—"the crank."

Fletcher, Joseph. 1988. *The Ethics of Genetic Control: Ending Reproductive Roulette.* Buffalo, NY: Prometheus Books.

This book unequivocally demonstrates the thesis that "ethics" is a codeword for genetic engineering.

Gardner, Martin, (ed.) 1965. *The Annotated Alice: "Alice's Adventures in Wonderland" and "Through the Looking Glass" by Lewis Carroll.* Harmondsworth: Penguin Books.

Alice is important to the technocracy. This annotated edition (first published in 1960) is by an editor of *Scientific American.*

Gilder, George. 1984. *The Spirit of Enterprise.* New York: Simon and Schuster.

An updated version of Nebuchadnezzar's Dream for the Reagan years.

Gowlett, John. 1984. *Ascent to Civilization: The Archaeology of Early Man.* New York: Alfred A. Knopf.

Who says they don't write archaeology like this any more?

Gribbin, John. 1985. *In Search of the Double Helix.* New York: McGraw-Hill.

This book, published in the Bantam New Age series, argues that quantum physics and biology are complementary disciplines. It straddles the boundary between conventional science and the emerging biological eschatology—but does not quite cross over.

Hall, Stephen S. 1989. "Professor Thorne's time machine," *California*, October: 68–75, 158–162.

A good example of the scientific popularization as overt theology.

Harris, Marvin. 1974. *Cows, Pigs, Wars & Witches: The Riddles of Culture.* New York: Random House.

In the works by this author, the findings of anthropology are assimilated to the technocratic folk model.

Johanson, Donald, and Maitland A. Edey. 1981. *Lucy: The Beginnings of Humankind.* New York: Simon and Schuster.

Human evolution assimilated to Nebuchadnezzar's Dream and explained by reductionist theories.

La Mettrie, Julien Offray de. 1912. *Man a Machine.* La Salle, IL: Open Court.

In this technocratic classic, first published in 1748, the mechanistic concept of the human body replaces philosophy and religion.

Landes, David S. 1983. *Revolution in Time: Clocks and the Making of the Modern World.* Cambridge, MA: Harvard University Press.

This meticulous history of clockwork from the 13th century to the present illustrates how modern scholarship rewrites history as Nebuchadnezzar's Dream. By equating "the clock" with a specific type of mechanism, the author can declare it to be a European invention and explicitly dismiss the Chinese contribution as a "dead end."

Laurence, William L. 1959. *Men and Atoms: The Discovery, the Uses, and the Future of Atomic Energy.* New York: Simon and Schuster.

This book by a former *New York Times* science editor contains first-hand descriptions of Trinity and Nagasaki reprinted from his earlier work, *Dawn Over Zero* (1946). Every significant technocratic image is presented here in vivid, gee-whiz prose: brilliant scientists, new dawns, stages of history, the Second Coming of Christ, the Secret of the Atomic Bomb, the Slavic betrayal, Nature unveiled, Alice in Wonderland, and many more.

Libby, Willard F. 1965. Man's place in the physical universe. In *New Views on the Nature of Man.* Edited by John R. Platt. Chicago: University of Chicago Press; pp. 1–15.

This Nobel laureate and inventor of radiocarbon dating reveals his dreams for the future—a world dependent on atomic power, in which the state controls reproduction through sperm banks.

Lumsden, Charles J., and Edward O. Wilson. 1983. *Promethean Fire: Reflections on the Origin of Mind.* Cambridge, MA: Harvard University Press.

Now Man can take control of the fourth stage of evolution—Mind! If this new Freedom makes you anxious, you can put your mind at rest with Scientific Ethics.

Morgan, Lewis Henry. 1985. *Ancient Society.* Tucson: University of Arizona Press.

First published in 1877, this work is the classic presentation of Nebuchadnezzar's Dream using anthropological materials.

Muller, H. J. 1973. *Man's Future Birthright: Essays on Science and Humanity.* Albany: State University of New York Press.

A Nobel laureate in genetics tells us what he really thinks.

Muller, Richard A. 1988. *Nemesis: The Death Star.* N.p.: Weidenfeld and Nicholson.

This is a good example of the fluid boundaries between science and sci-fi: the author argues on astronomical grounds that our sun has an invisible twin.

Nebelsick, Harold. 1981. *Theology and Science in Mutual Modification.* New York: Oxford University Press.

This progressive theologian thinks modern science is an open-minded inquiry into the laws of nature.

Neufeldt, Victoria, and David B. Guralnik, (eds.) 1988. *Webster's New World Dictionary of American English. Third College Edition.* New York: Simon and Schuster.

A technocratic definition of the English language, complete with canonical etymologies.

Paul, Iain. 1982. *Science, Theology and Einstein.* New York: Oxford University Press.

This theologian thinks that Einstein is a source of spiritual inspiration.

Renfrew, Colin. 1973. *Before Civilization: The Radiocarbon Revolution and Prehistoric Europe.* London: Cambridge University Press.

The assimilation of archaeological findings to the absolute chronology of physics.

Rhodes, Richard. 1988. *The Making of the Atomic Bomb.* New York: Simon and Schuster.

This is the canonical technocratic account of the Manhattan Project. Revealingly, the author says he used only printed sources because the memories of the scientists he interviewed were too unreliable.

Rhodes, Richard. 1990. *A Hole in the World: An American Boyhood.* New York: Simon and Schuster.

In his autobiography the author of *The Making of the Atomic Bomb* assimilates his own life to the myth of The Fire Twins.

Rorvik, David. 1971. *Brave New Baby: Promise and Peril in the Biological Revolution.* Garden City, NY: Doubleday.

Science journalism by a believer in test-tube babies.

Rorvik, David M. 1978. *In His Image: The Cloning of a Man.* Philadelphia: J.B. Lippincott Co.

A revealing work of fiction that sets out the agenda of genetic engineering.

Schrödinger, Erwin. 1967. *What Is Life? and Mind and Matter.* Cambridge: Cambridge University Press.

What Is Life?, first published in 1944, provides a physicalist charter for would-be molecular biologists.

Seaborg, Glenn T. 1972. *Nuclear Milestones: A Collection of Speeches by Glenn T. Seaborg.* San Francisco: W.H. Freeman.

The technocratic agenda of Big Science presented to a scientific and technical audience.

Seaborg, Glenn T. and William R. Corliss. 1971. *Man and Atom: Building a New World Through Nuclear Technology.* New York: E. P. Dutton & Co.

A classic work of technocratic boosterism by a major player.

Service, Elman R. 1962. *Primitive Social Organization: An Evolutionary Perspective.* New York: Random House.

The assimilation of the findings of social anthropology to Nebuchadnezzar's Dream.

Smiles, Samuel. 1966. *Selections From The Lives of the Engineers: An Account of Their Principal Works.* Cambridge, MA: M. I. T. Press.

These selections, first published in the 19th century, assimilate engineering to the myth of The Lone Galileo.

Smith, Martin Cruz. 1986. *Stallion Gate.* New York: Random House.

This thriller reinforces the notion that the Manhattan Project was in danger from enemy agents; the title references the main (northern) gate to the Trinity test site.

Stableford, Brian. 1985. *Future Man.* New York: Crown Publishers.

This work, first published in England in 1984, presents the future possibilities of genetic engineering in such graphic and overt form that the totalitarian premises are completely transparent. Naturally, it is hard to get.

Stark, D. E. 1990. "Designer genes: the ethics of biotechnology," *High Technology Careers Magazine,* Vol. 7, No. 3, June/July: pp. 16, 19.

A good example of how the social consequences of biotechnology are always presented in the non-scientific press as a problem in the "ethical" use of a scientific fait accompli.

Suzuki, David, and Peter Knudson. 1990. *Genethics: The Ethics of Engineering Life.* Cambridge, MA: Harvard University Press.

Genetics and ethics are now completely conflated—as "genethics."

Thompson, William Irwin. 1989. *Imaginary Landscape: Making Worlds of Myth and Science.* New York: St. Martin's Press.

This book is a lead-in to the technocratic theocracy: mathematics as spirituality, Gaia as global unification, AIDS as a positive evolutionary force, etc.

Vinge, Joan D. 1983. *Star Wars: Return of the Jedi.* New York: Random House.

A spin-off book from the *Star Wars* film.

Wade, Nicholas. 1977. *The Ultimate Experiment: Man-Made Evolution.* New York: Walker & Co.

The senior staff writer for *Science* tells us that recombinant DNA is the first step in making genetic engineering technically feasible. Will *Homo sapiens* have the courage to "bring to birth his finest creation: *Homo sapientissimus*" (p. 7)? As long as men care about Freedom and Progress, he will.

Watson, James D. 1980. *The Double Helix: A Personal Account of the Discovery of the Structure of DNA. With text, commentary, reviews, and original papers, edited by Gunther S. Stent.* New York: Norton & Company.

First published in 1968, it claims to strip science of its noble facade and humanistic claims—thereby reducing it to technocracy.

Wells, H. G. 1914. *The World Set Free.* New York: E. P. Dutton & Co.

This forgotten work presents the ritual cycle for the Manhattan Project and molecular biology.

Wells, H. G. 1923. *Men Like Gods.* London: Cassell and Company.

This forgotten work connects the myth of progress with the absence of an immune system.

Wells, H.G. 1975. *Early Writings in Science and Science Fiction.* Berkeley and Los Angeles: University of California Press.

The cover art shows a disembodied head and hands—a perfect icon for the contents.

Wilson, E. O. 1978. *On Human Nature.* Cambridge, MA: Harvard University Press.

The inventor of sociobiology speaks with surprising candor about the atheological agenda of science.

Wingerson, Lois. 1990. *Mapping Our Genes: The Genome Project and the Future of Medicine.* New York: Dutton.

A well-written and well-meaning account of the Human Genome Initiative by a journalist who believes that medical science is primarily concerned with the healing of sick people.

Wouk, Herman. 1973. *The Winds of War.* New York: Pocket Books.

First published in 1971, this two-volume novel was made into a television miniseries in the late 1980's. It exemplifies the myth of The Nazi/ the Jew/ and the Yank.

Wright, Robert. 1988. *Three Scientists and Their Gods: Looking for Meaning in an Age of Information.* New York: Times Books.

A wide-eyed account of the technocratic future based on interviews with players.

TECHNOCRATIC LEXICON

A list of words, proper names, letters, and numerals that have a significant symbolic role in the surface structure of contemporary technocratic mythology.

advance
AIDS
Alamogordo, New Mexico
Alice in Wonderland
alpha
America
analyze
arms, both limbs and weapons
arrow
Athens, Greece
Atlantis
atom
atomic bomb
Bacon, Francis
Berkeley, University of California at
Big Bang
black
Bohr, Niels
brain
breakthrough
California
Caltech (California Institute of Technology)
Campanile
carbon
catapult
caterpillar
Center for Advanced Studies
chain reaction
civilization
clock
cocoon
Cold Spring Harbor Laboratory
Columbia, the University and the space shuttle
comet
computer
Congo
Copernicus
copper
crank
Crick, Francis
cryogenic
crystal
Curie, Madame Marie
cylinder
Da Vinci, Leonardo
Darwin, Charles
diamonds
digit, both fingers and numbers
discover
DNA
Donne, John
Double Helix
Drosophila
Egypt
Einstein, Albert
ethics
explore
father
Fermi, Enrico
Fermilab
fireflies
fission
Florence, Italy
forward
freedom
fusion
galaxy
Galileo
gene
genetics
Germany
glacier
global
glove
gold
Hanford, Washington
Harvard University
head
Heisenberg, Werner
helium
helix
hemisphere
high
Hiroshima
Hitler, Adolf
Holocaust
immune
Indian (American)
industrial
intercontinental
intravenous
iron
Japan
Johns Hopkins University
Jornada del Muerto
July 16
Jupiter, the planet and the god
Kepler, Johannes
Kissinger, Henry
Kodak Corporation

INDEX

CREDITS

Book Development and Production:
System Translation, Inc., Palo Alto, CA

Illustration: Erica Aitken

Book Design: Helga Wild

Typesetting: Byron Brown

Publication consultants: Mary Douglas and Mitchell Allen

Copyediting: Louise Herndon

Prepublication review:
Myrdene Anderson, Purdue University
Niklas S. Damiris, Xerox PARC
Candace L. Holts, System Translation, Inc.
Diane Jonte-Pace, Santa Clara University
Nicole Sault, Santa Clara University
Richard Sonnenfeld, IBM Corporation
Helga Wild, Iconic Anthropology Press

ABOUT THE AUTHOR

Peter C. Reynolds is an anthropologist who left the groves of academe to seek his fortune in Silicon Valley. From 1984 to the present, he has worked with high-tech organizations in various capacities, including software instructor, technical education manager, and now as the owner of his own consulting firm. In addition to his computer industry experience, Mr. Reynolds has extensive academic qualifications. After receiving his Ph.D. degree in anthropology from Yale University in 1972, he became Postdoctoral Fellow at Stanford University Medical Center, where he did research on the neuropsychology of cognitive processes. From 1974 to 1980 he was Research Fellow at Australian National University in Canberra, Australia. He did fieldwork in Malaysia, Australia, and Papua New Guinea; and in 1980 he received a Harry Frank Guggenheim Foundation grant to do a comparative study of technical skills using video technology. Before completing his doctoral degree, Mr. Reynolds worked at the Salk Institute for Biological Studies, where he was research assistant to J. Bronowski, creator of the BBC television series *The Ascent of Man*. At Yale, he was assistant to Theodosius Dobzhansky, the well-known geneticist. Mr. Reynolds is the author of many articles and reviews, as well as the book *On the Evolution of Human Behavior*, published by the University of California Press in 1981.

To Order Copies of This Book

Order by telephone:

If you are calling from the United States and have a current MasterCard or Visa card, then call our 24-hour, toll-free, voice-mail number:

1 800 456 5131

The voice mail system will ask you for the type of card (Visa or MasterCard), your name as it appears on the credit card, the card number, the expiration date, your address, the title of the book, and the number of copies requested.

Order by mail:

Mail orders may be paid by money order or by a check drawn on an American bank in U. S. dollars. If the order is shipped to an address outside the United States, there is an additional handling fee.

For prompt service, enclose a facsimile of the form shown on the next page. Make checks and money orders payable to Iconic Anthropology Inc. Send your order to:

Order Department
Iconic Anthropology Press
Post Office Box 50217
Palo Alto, CA 94303
USA

Additional information:

For information on quantity discounts (10 or more copies), distribution policy, and rush shipping options, call:

1 800 456 5131

Order Form for Stealing Fire

Cover price of *Stealing Fire*	$ 24.95	
If the order is to be shipped outside the U. S. or Canada, add a $5.00 handling charge:	$ 5.00	*International only!*
Unit cost: Cover price plus handling charge (if any)	$ _____	
Number of copies requested _____		
Order Subtotal: Number of copies times unit cost:	$ _____	
If the order is to be shipped to a California address, add applicable sales tax.	$ _____	*California only!*
Order Total: Subtotal plus sales tax (if any):	$ _____	

Write a check or money order for this amount, payable to Iconic Anthropology Inc. Mail your payment and order form to:

Order Department
Iconic Anthropology Press
Post Office Box 50217
Palo Alto, CA 94303

Prices are subject to change without notice. All amounts are in United States dollars. Checks must be drawn on an American bank in U. S. dollars. Orders are shipped by mail unless other arrangements have been made in advance.

Order Form for Stealing Fire

Cover price of *Stealing Fire*	$ 24.95	
If the order is to be shipped outside the U. S. or Canada, add a $5.00 handling charge:	$ 5.00	*International only!*
Unit cost: Cover price plus handling charge (if any)	$ _____	
Number of copies requested _____		
Order Subtotal: Number of copies times unit cost:	$ _____	
If the order is to be shipped to a California address, add applicable sales tax.	$ _____	*California only!*
Order Total: Subtotal plus sales tax (if any):	$ _____	

Write a check or money order for this amount, payable to Iconic Anthropology Inc. Mail your payment and order form to:

Order Department
Iconic Anthropology Press
Post Office Box 50217
Palo Alto, CA 94303

Prices are subject to change without notice. All amounts are in United States dollars. Checks must be drawn on an American bank in U. S. dollars. Orders are shipped by mail unless other arrangements have been made in advance.